The Way To Pass

Maths

Level 6

GW00501693

The Way to Pass
National Curriculum
Maths
Level 6

Geoff Buckwell

VERMILION
LONDON

First published in 1994

1 3 5 7 9 10 8 6 4 2

Text copyright © Rockhopper 1994

The moral right of the Author has been asserted
in accordance with the Copyright, Designs and
Patents Act, 1988.

All rights reserved. No part of this publication
may be reproduced, stored in a retrieval system,
or transmitted in any form or by any means,
electronic, mechanical, photocopying, recording
or otherwise, without the prior permission of
the copyright owner.

First published in the United Kingdom in 1994
by Vermilion
an imprint of Ebury Press
Random House, 20 Vauxhall Bridge Road,
London SW1V 2SA

Random House Australia (Pty) Limited
20 Alfred Street, Milsons Point, Sydney,
New South Wales 2061, Australia

Random House New Zealand Limited
18 Poland Road, Glenfield,
Auckland 10, New Zealand

Random House South Africa (Pty) Limited
PO Box 337, Bergvlei, South Africa

Random House UK Limited Reg. No. 954009

Editor: Alison Wormleighton
Design: Jerry Goldie Graphic Design

A CIP catalogue record for this book
is available from the British Library

ISBN 0-09-178125-6

Typeset by AFS Image Setters Ltd, Glasgow
Printed in Great Britain by Butler & Tanner Ltd,
London and Frome

Foreword

Welcome to THE WAY TO PASS NATIONAL CURRICULUM MATHS LEVEL 6. I want to tell you why I have put together this series of books, along with a team of teachers, advisers and examiners.

A lot of people don't enjoy Maths because they're frightened of it. I can understand how frightening it can be because I've been scared of it myself at times: maybe the teacher goes through the work a little too quickly for you, maybe there are too many children in your class, maybe you're not the best at Maths in your class. All of these reasons can make Maths seem impossible. What I have learned through the years is that the more help you have and the longer you spend on something, the more likely you are to get over any difficulties.

Whatever you might think about school, and about Maths in particular, there is no doubt that Maths and English are the two most important subjects for you to do well in. If you understand most of what you're taught, you are set for a brighter future, being able to do some of the things you've always wanted to. The WAY TO PASS series can help you through secondary school, making the subjects you're taught a little more understandable and interesting, making your tests easier and helping you to get the best grades possible.

All of the books are based around work for you to do at home. Most of the explanations will have been covered in classes at school and so you won't want to wade through pages and pages of more explanations. That is why in each section we give you a concise list of the main things you need to know, and then work through exercises to practise each one.

This completely new range of books has been organised so that, if you want to, you can follow the already successful VIDEO CLASS videos covering the same subjects/levels. All of the book sections work together neatly with the video sections so that you have a complete course at your fingertips. Alternatively, the books can be used on their own, without the videos.

I certainly hope that this series will make Maths and English more approachable and slightly friendlier than they were before. Remember, you must follow what is taught in school and do as many exercises as you can—the more practice you get, the better you will be.

Carol Vorderman

Contents

The National Curriculum

The National Curriculum sets targets for pupils of all abilities from age 5 to 16, specifying what they should know, understand and be able to do at each stage of their education. It is divided into four **Key Stages**: Key Stage 1 (age 5–7), Key Stage 2 (age 7–11), Key Stage 3 (age 11–14) and Key Stage 4 (age 14–16).

The GCSE examinations are the main way of assessing children's progress at the end of Key Stage 4 (age 16). Prior to that, at the end of Key Stages 1, 2 and 3 (i.e. at ages 7, 11 and 14), pupils will be assessed in two ways: continuous assessment by the teachers and national tests, in which children will be asked to perform specific tasks relevant to the subject. Children will take the tests in English and Maths at age 7, 11 and 14 and in Science at age 11 and 14. These three subjects are at the heart of the National Curriculum and are known as the **core subjects**.

By combining the test results and continuous assessment, a teacher will be able to determine the **Level** a child has reached in each of these subjects. Different children at the same Key Stage may achieve widely varying results and therefore different Levels.

An average child will probably move up one Level every two years or so, starting at Level 1 at the age of 5. This means that at the end of Key Stage 1 (age 7) they may reach Level 2, at the end of Key Stage 2 (age 11) Level 4 and at the end of Key Stage 3 (age 14) Level 5 or 6. Slower learners could be a Level or two lower in one or more subjects, while some children could be two or even three Levels higher. Level 10 is the highest, but only a few children will achieve this Level.

The books for Levels 4, 5 and 6 in THE WAY TO PASS series are based on National Curriculum requirements for each of those Levels and are suitable for the secondary school child aged 11 to 14. They will serve as a valuable back-up to a child's classwork and homework and provide an excellent preparation for the tests at the end of Key Stage 3.

Introduction

The WAY TO PASS books for Key Stage 3 are written to provide you with that extra support you will need at home while preparing for the National Curriculum tests. This book is aimed at Maths Level 6 pupils. (If you are not sure which Level you are working towards, your teacher will be able to tell you.)

Part of your revision will be learning the facts, which are listed under **Things You Need to Know** at the beginning of each section in the book. But the largest part of your time should be spent actually working out problems. In **How to Do It** in each section there are worked examples. As you do these, cover up the solutions to check you understand that particular topic. When you feel confident, try the **Do It Yourself** questions in that section. You'll find the **Answers** near the end of the book.

The questions marked with a picture of a calculator with a cross through it must be done without a calculator. (It is extremely important in the real tests that you show your full working – including carry digits – in this type of question; failure to do so would mean you'd score no marks for that question.) Where a question has a picture of a calculator with no cross, a calculator *should* be used.

The numbering system used in the book makes it easy for you to concentrate on whatever topics you feel you most need to revise. Each topic within a section has a number, which identifies that topic throughout the section. Thus, in Section 1, for example, an explanation of powers appears in no. 7 of Things You Need to Know; then exercise 7 of How to Do It shows you how to answer questions involving powers; and finally you can check how well you understand powers with exercises 7a, 7b and 7c of Do It Yourself.

Many questions in the National Curriculum tests will cover a range of topics, and you should pay particular attention in this book to the work in section 11 on shape and in section 13 on algebra.

At the end of the book there is a **Sample Test Paper** based on the type of questions you will be given, followed by solutions. It should take you about 60 minutes to do. If this were a real Key Stage 3 test, you would need to get about three-quarters of the answers correct to be successful at Level 6.

Don't leave your revision until the last minute before your Key Stage 3 test. Remember, the more you practise, the better you will cope with it.

1 | Working with Numbers

Things You Need to Know

1 How to multiply by long multiplication without using a calculator.

2 How to divide using a 'pencil and paper' technique without the use of a calculator.

3 How to multiply and divide by 10, 100, etc. without the use of a calculator.

4 A **factor** of a number divides exactly into that number. For example, 8 is a factor of 24, but 5 is not.

A **prime** number has no factors other than 1 and the number itself. The first few prime numbers are 2, 3, 5, 7, 11, . . . (Notice that 1 is not a prime number.)

The **multiples** of 6, for example, are 6, 12, 18, 24, 30, . . . (the 'six times table').

5 **Negative** numbers are numbers that are less than zero. So $8 - 11 = -3$ (we say either 'minus 3' or 'negative 3'). The rules for multiplying and dividing are given in the table shown alongside:

× or ÷	Positive	Negative
Positive	Positive	Negative
Negative	Negative	Positive

6 A decimal is another way of writing a fraction using $\frac{1}{10}$, $\frac{1}{100}$, $\frac{1}{1000}$, etc. So 0.436 means

$$\frac{4}{10} + \frac{3}{100} + \frac{6}{1000} \quad \text{or} \quad \frac{436}{1000}$$

You must be able to arrange decimals in order of size.

7 **Powers** or **index** numbers are used to abbreviate things like $4 \times 4 \times 4 \times 4 \times 4$ which can be written as 4^5 (four to the power 5).

> *Note:* (i) Power 2 is the same as squared.
>
> Power 3 is the same as cubed.
>
> (ii) $6^1 = 6$, $6^0 = 1$ (any number to the power zero is 1).

The reverse of squaring is square rooting; so

$$\sqrt{25} = 5 \quad \text{because } 5^2 = 25$$

The reverse of cubing is cube rooting; so

$$\sqrt[3]{8} = 2 \quad \text{because } 2^3 = 8$$

11

How to Do It

1 Work out 326 × 23, showing all your working.

You must show all carry digits

Solution

Set out the multiplication with the smaller number underneath, making sure the digit columns are in line.

$$
\begin{array}{r}
3\ \ 2\ \ 6 \\
\times\ \quad 2\ \ 3 \\
\hline
9\ \ 7,\ 8 \\
6\ \ 5,\ 2\ \ 0 \\
\hline
7,\ 4\ \ 9\ \ 8
\end{array}
$$

    (i) multiply by 3

    (ii) multiply by 20

Hence 326 × 23 = 7498.

There are many ways of multiplying and dividing, and, if you already use methods that you can remember easily and that give correct answers, then do stay with them! If you are uncertain about any of it, however, the methods shown here will be useful.

2 Work out 276 ÷ 12, showing all your working.

Solution

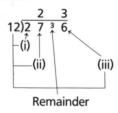

Remainder

(i) 12 does not divide into 2

(ii) 12 divides into 27 twice, leaving a remainder 3

(iii) 12 divides into 36 three times

Hence 276 ÷ 12 = 23.

As you will find when using a calculator, multiplying and dividing by 10, 100 or 1000 is very easy. When you watch the calculator display it is as if the original number moves to the left in the calculator display and becomes larger (in the case of multiplication) or moves to the right and becomes smaller (in the case of division). This means that it is very easy to do such calculations without a calculator. It really amounts to deciding where to place the decimal point.

3 Carry out the following multiplications and divisions without the use of a calculator:

(i) 860×100 (ii) 23×200 (iii) $460\,000 \div 100$

(iv) $84 \div 100$ (v) $640 \div 800$

Solution

(i) 860 can be written 860.00. $\times 100$ will move the decimal point two places to the right to become 86 000. So

$$860 \times 100 = 86\,000$$

(ii) First multiply $23 \times 2 = 46$. Then multiply 46 by 100 as in part (i). So

$$23 \times 200 = 4600$$

(iii) Since 460 000 is really 460 000.0, if we divide by 1000 the decimal point moves three places to the left to become 460.000. So

$$460\,000 \div 1000 = 460$$

(iv) 84 can be written 84.0. To divide by 100 move the decimal point two places to the left to become 0.84. So

$$84 \div 100 = 0.84$$

(v) First divide 640 by 8 to give 80. Then divide by 100 as in part (iv). So

$$640 \div 800 = 0.80 \text{ or just } 0.8$$

4 This question is about the number 24.

(i) List the factors of 24.

(ii) Which of these factors are prime numbers?

(iii) Multiply together your answers in part (ii). What do you notice about the answer?

Solution

(i) The factors of 24 are 1, 2, 3, 4, 6, 8, 12, 24.

(ii) 2 and 3 are the prime factors.

(iii) $2 \times 3 = 6$, which is also a factor of 24.

'A prime factor is a factor which also happens to be a prime number'

'Remember to include 1 and 24 as factors of 24'

*'The number
line is very
useful when
working with
negative
numbers'*

5 **a** Carl belonged to the school weather club. On December 4th, he took the temperature at midnight. It was $-2\,°C$. The following morning at midday, it had risen to $8\,°C$. How much had the temperature changed in 12 hours?

Solution

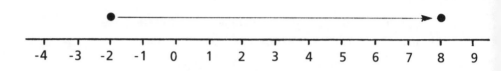

You can see from the diagram that the increase is $8° + 2° = 10\,°C$.

b Tim and Jason are playing a board game in which they are not allowed to use a calculator. They draw a card which has a sum on it. If the answer is positive they move that number of squares forward; if it is negative, they move that number of squares backwards. Jason has three goes and draws the following cards:

$$\boxed{-3+6} \qquad \boxed{-24 \div -3} \qquad \boxed{-2 \times 2}$$

How many squares does he move altogether, and in which direction?

Solution

$$-3 + \ 6 = +3$$
$$-24 \div -3 = +8$$
$$-2 \times \ 2 = -4$$

The total number of squares he moves will be

$$+3 + (+8) + (-4) = +7$$

Hence he moves 7 squares forward.

c Use your calculator to work out -1.2×-8.4.

Solution

Most calculators have a button $\boxed{+/-}$ for entering negative numbers.

This is pressed after the number. So

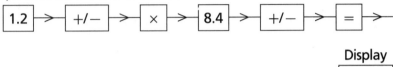

Display

$\boxed{10.08}$

Hence $-1.2 \times -8.4 = 10.08$.

'Read the instruction book with your calculator'

6 a Express the following decimals as fractions as simply as possible:
   (i) 0.32   (ii) 0.375

Solution

(i)

0.32 means $\dfrac{32}{100}$

Cancel top and bottom by 4

$\dfrac{\cancel{32}^{\,8}}{\cancel{100}_{\,25}} = \dfrac{8}{25}$

(ii)

0.375 means $\dfrac{375}{1000}$

You may not be able to see how to cancel this fraction in one step.

First cancel by 5 $\dfrac{\cancel{375}^{\,75}}{\cancel{1000}_{\,200}} = \dfrac{75}{200}$

Then cancel by 5 again $\dfrac{\cancel{75}^{\,15}}{\cancel{200}_{\,40}} = \dfrac{15}{40}$

Finally, cancel by 5 for a third time $\dfrac{\cancel{15}^{\,3}}{\cancel{40}_{\,8}} = \dfrac{3}{8}$

'You could cancel once by 125'

b Arrange the following decimals in order of size, smallest first: 1.6, 2.027, 0.84, 2.33.

Solution
Look at the number to the left of the decimal point first; as 2.027 and 2.33 both begin with 2, look to the right of the decimal point, where 0.33 is larger than 0.027 (take no notice of how many decimal places there are). Continuing this approach, the order is

0.84, 1.6, 2.027, 2.33

7 Use a calculator to work out 8^5.

Solution
Method 1
If you do not have a scientific calculator, you will have to use the $\boxed{\times}$ key four times.

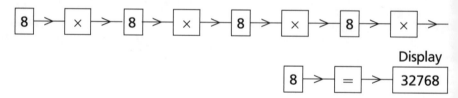

Method 2
If you have a scientific calculator, you will have a power button $\boxed{x^y}$. So

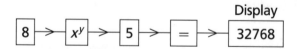

In either case, $8^5 = 32\,768$.

Do It Yourself

Answers are given in the Answers section at the back of the book.

1 Work out, showing all your working:

(i) 124×12 (ii) 86×35 (iii) 643×29

2 Find, without using a calculator:

(i) $396 \div 11$ (ii) $2016 \div 18$ (iii) $25\,473 \div 21$

3a Write down the value of:

(i) 30×50 (ii) $8600 \div 20$ (iii) $70\,000 \div 200$
(iv) $28 \div 200$ (v) $490 \div 700$ (vi) 3.6×2000

b Tina has decided to plant some sunflower seeds for charity. One packet contains 12 seeds. Unfortunately, she does not have her calculator and wants to know how many seeds she will have if she buys 16 packets. Show your working to find out how many she will get.

In a competition, you pay 20p a seed to enter. If she sold 320 packets, how much money would she raise?

4a List all factors of the following numbers:

(i) 20   (ii) 40   (iii) 60   (iv) 80

b Find the smallest number that 8, 10 and 12 all divide into exactly.

5a | Start | → | Add -5 | → | Subtract 3 | → | Divide by -2 | → | Finish |

(i) If you start with the number 20, and carry out the above instructions, what do you finish with?
(ii) If you finished with 8, what did you start with?

b Work out:

(i) $8 + -6$ (ii) -3×-12 (iii) $-6 - -5$
(iv) $-2\frac{1}{2} + 6\frac{1}{4}$ (v) $15 \div -2$ (vi) $(4-8) \times (3-6)$

c Pick out three cards from the following, so that the difference between the numbers on two of them equals the number on the third.

Can you pick out a *different* selection of three cards for which the same property is true?

6 **a** In the number 834.6, the 3 really stands for 30. What does the 3 stand for in the following numbers:

(i) 348 (ii) 103.6 (iii) 8.03 (iv) 0.346

b Change the following decimals into fractions, giving your answer in its simplest form:

(i) 0.35 (ii) 0.675 (iii) 0.84

7 **a** Work out:

(i) 2^4 (ii) 3^5 (iii) 8^0 (iv) 10^4 (v) $\sqrt{81}$ (vi) $\sqrt[3]{729}$

b Use a calculator to work out:

(i) 4^6 (ii) 2^9 (iii) 3^7 (iv) 1.2^3 (v) 0.24^6 (vi) 2.04^5

c Sean has said to his friend Peter that he knows how to win at roulette. He puts £2 on the colour black. If he loses, he then puts £2 × 2 = £4 on black. If he loses again, he then puts £2 × 2 × 2 = £8 on black. If he loses 12 times, how much will he then have to put on the black at the next turn? Write your answer as a power of 2. Then use a calculator to work out the actual amount.

Number Patterns

Things You Need to Know

1 A **number pattern** is a sequence of numbers in which it is possible to predict the next number from previous numbers in the pattern.

Certain number sequences have special names:

(i) 1, 2, 3, 4, ...   natural numbers
(ii) 2, 4, 6, 8, ...   even numbers
(iii) 1, 3, 5, 7, ...   odd numbers
(iv) 1, 4, 9, 16, ...   square numbers

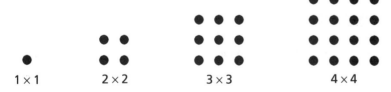

(v) 1, 3, 6, 10, ...   triangular numbers

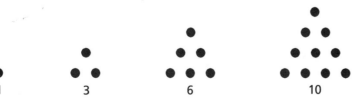

'Dot patterns are often helpful'

2 The **difference** between successive numbers in a number pattern often helps you see the pattern more clearly. Look at the triangular numbers:

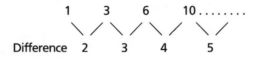

The next number in the sequence is $10 + 5 = 15$.

Numbers can sometimes be predicted in a number pattern by a rule expressed in symbols (a formula).

3 How to generate a number sequence using a simple computer program.

How to Do It

1 Find the missing numbers in the following number patterns:

(i) 4, 7, 10, 13, 16, ☐, ☐, . . .

(ii) 20, 16, 12, 8, 4, ☐, ☐, . . .

(iii)
```
                1
             1     1
          1     2     1
       1    ☐    3    1
    1    4   ☐   4    1
```

(iv) $\frac{1}{2}, \frac{1}{4}, \frac{1}{8}, \frac{1}{16},$ ☐, ☐, . . .

Solution

(i) Each number is 3 greater than the previous number. The next two numbers are 19, 22.

(ii) Each number is 4 less than the previous number. The next two numbers are 0, -4.

(iii) Each number in the triangle is the sum of the two numbers above it. The completed triangle looks as follows:

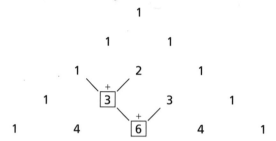

(iv) The denominator is multiplied by 2 each time. The next two are $\frac{1}{32}$, $\frac{1}{64}$.

2a Look at this table:

	Number of dots D	Number of lines L
	2	1
	3	2
	4	3
	5	4

The diagram below shows how you can work out the number of lines when you know the number of dots.

Number of dots → Subtract 1 → Number of lines

Write the rule for this using D and L.

Solution
Mathematicians express rules by means of formulae. Since it is to find L, you start with $L = $. So

$$L = D - 1$$

'*Check that your formula works for the numbers given*'

b The diagram shows a series of matchstick patterns. Write down the number of matches in each diagram, and predict how many matches would be needed for the next pattern. Explain carefully how you got your answers.

  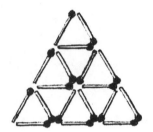

Solution
By counting, we find the number of matches in each shape is as follows:

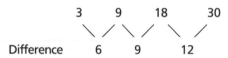

$$3 \qquad 9 \qquad 18 \qquad 30$$

Difference   6   9   12

In the pattern of differences, the next difference will be 15. Hence the next pattern will have $30 + 15 = 45$ matches.

c

X	C
0	4
1	8
4	12
9	16

The patterns above consist of circles (*C*) and crosses (*X*). The table on the left shows how many of each are used:

(i) The rule that connects *C* and *X* is as follows:

$$C \rightarrow \boxed{\div 4} \rightarrow \boxed{\text{Subtract 1}} \rightarrow \boxed{\text{Square}} \rightarrow X$$

Find *X* if *C* = 40.
(ii) What do you notice about *C* + *X*?

*'A flow chart
can be used
instead of a
formula'*

Solution

(i) Following the rule, we have

$$40 \div 4 = 10$$
$$10 - 1 = 9$$
$$X = 9^2 = 81$$

(ii) The answers for $C + X$ are as follows:

$C + X$: 4, 9, 16, 25

The answers are all square numbers.

3 Investigate the number sequence generated by the following computer program:

```
10  FOR NUMBER = 1 TO 8
20  PRINT 2*NUMBER + 1
30  NEXT NUMBER
40  END
```

'Try this and similar programs on a computer'

Solution

Lines 10 and 30 tell you to repeat a calculation, with NUMBER = 1, 2, 3, 4, 5, 6, 7, 8 (this is called a loop).

Line 20 tells you to work out

$$2 \times 1 + 1 = 3$$
$$2 \times 2 + 1 = 5$$
$$\vdots \qquad \vdots \qquad \vdots \qquad \vdots$$
$$2 \times 8 + 1 = 17$$

'Skip this if you have not yet met BASIC programming'

The number sequence you get is

3, 5, 7, 9, 11, 13, 15, 17

Hence the program gives us the odd numbers from 3 to 17 inclusive.

Do It Yourself

Answers are given in the Answers section at the back of the book.

1 Fill in the gaps in the following number patterns:

(i) 1, 2, 4, 7, 11, ☐, ☐, . . .

(ii) 1, 1, 2, 3, 5, ☐, ☐, . . .

(iii) 2, 5, 10, 17, ☐, ☐, . . .

(iv) $\frac{1}{2}, \frac{3}{4}, \frac{7}{8}, \frac{15}{16}$, ☐, ☐, . . .

2 The diagram below shows a sequence of patterns of squares made from metal links. The number of squares S is related to the number of links L by the rule $L = 3S + 1$.

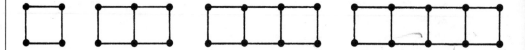

(i) Use the rule to predict the number of links needed for the next two shapes.

(ii) How many links would you need to make the twentieth shape in the pattern?

(iii) If you had 60 links, would you be able to use all of them to make one of the shapes in your pattern? Explain your answer.

'Leave 3a and b out if you have not yet met BASIC programming'

3a Investigate the number sequence generated by the following computer programs:

(i)
```
 5  FOR N = 1 TO 8
10  PRINT N*N+1
15  NEXT N
20  END
```

(ii)
```
20  FOR N = 1 TO 6
30  PRINT 5*N−3
40  NEXT N
50  END
```

b Write a simple program to generate each of the following number sequences:

(i) 3, 5, 7, 9, 11, 13, 15

(ii) 1, 4, 9, 16, 25, 36

(iii) 1, 5, 9, 13, 17, 21

(iv) 1, 8, 27, 64, 125

Fractions, Percentages and Ratios

Things You Need to Know

1 A fraction is a *part* of something. It is written as

$$\frac{\text{numerator}}{\text{denominator}}$$

For example, in the fraction $\frac{2}{3}$, 2 is the **numerator**, and 3 is the **denominator**. (Any fraction can be called a **vulgar** fraction.)

Equivalent fractions are equal fractions, but written with different denominators. For example, $\frac{2}{3} = \frac{4}{6}$. (The numerator and denominator have both been multiplied by 2.)

An **improper** fraction is one in which the numerator is bigger than the denominator. (The top is bigger than the bottom.) For example, $\frac{9}{5}$ is an improper fraction.

A **mixed number** is one such as $4\frac{2}{3}$.

2 A fraction can be changed into a decimal by dividing the numerator by the denominator. So for $\frac{5}{8}$, find $5 \div 8 = 0.625$ (by calculator).

3 The **percentage** sign (%) means out of 100. So 40% means 40 out of 100, or as a fraction $\frac{40}{100}$.

You can change a fraction or decimal into a percentage by multiplying it by 100.

'A ratio is a way of talking in numbers about the relationship between two things'

4 A **ratio** such as 3:1 means that there is three times as much of one quantity as the other.

How to Do It

1 a Shade $\frac{5}{12}$ of this shape:

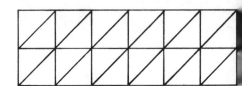

Solution

This question tests whether you understand equivalent fractions. If you look carefully, the diagram contains 24 triangles. This means the denominator of the fraction needs to be 24. So

$$\frac{5}{12} = \frac{5 \times 2}{12 \times 2} = \frac{10}{24} \longleftarrow \text{10 triangles need to be shaded}$$

The final answer is shown alongside. (You can shade any 10 triangles you like.)

b Change $3\frac{2}{3}$ into an improper fraction.

Solution

First represent the mixed number, using circles divided into thirds.

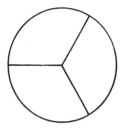 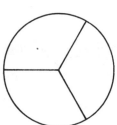 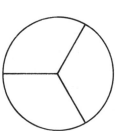 

'Do not confuse $3\frac{2}{3}$ with $\frac{32}{3}$'

Now count the total number of thirds in the diagram. You can see there are $3 \times 3 + 2 = 11$. So the answer is that $3\frac{2}{3} = \frac{11}{3}$.

You can leave out the diagram if you feel confident.

c Jason usually spends $\frac{3}{5}$ of his pocket money on sweets in a week. His mum gave him £1.60 on Monday. How much did he have left at the end of the week?

Solution
Method 1
To find $\frac{3}{5}$ of £1.60, you work out $\frac{3}{5} \times £1.60$:

$$\frac{3 \times £1.60}{5} = \frac{£4.80}{5} = £0.96$$

Hence Jason spends 96 pence.

At the end of the week, he has £$(1.60 - 0.96) = £0.64$ or 64 pence left.

Method 2
Change £1.60 to pence, and find $\frac{1}{5}$ by dividing by 5.

$$5\overline{)16^10} \quad 3\ 2$$

So $\frac{1}{5}$ is 32 pence. Now if he spends $\frac{3}{5}$, he must save $\frac{2}{5}$. The amount he has left is $2 \times 32 = 64$ pence.

2 Express the following decimals as fractions in their simplest form (lowest terms):
(i) 0.4   (ii) 0.45

Solution
(i) The 4 in the first decimal place means four-tenths. So

$$0.4 = \frac{4}{10}$$

Cancel top and bottom by 2

$$\frac{\overset{2}{\cancel{4}}}{\underset{5}{\cancel{10}}} = \frac{2}{5}$$

Hence

$$0.4 = \frac{2}{5}$$

If you have a calculator with a fraction button $\boxed{a^{b/c}}$ it will simplify a fraction for you as shown overleaf:

'Choose the method you prefer'

'Remember the first decimal place is $\frac{1}{10}$, the second is $\frac{1}{100}$ and so on'

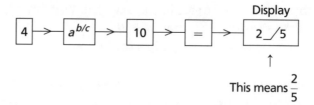

This means $\frac{2}{5}$

(ii) This contains hundredths, so

$$0.45 = \frac{45}{100}$$

Cancel top and bottom by 5

$$\frac{\overset{9}{\cancel{45}}}{\underset{20}{\cancel{100}}} = \frac{9}{20}$$

So

$$0.45 = \frac{9}{20}$$

3 **a** In a survey of 40 houses, 22 had a video recorder. What is the percentage of houses that had a video recorder?

Solution

The fraction that had a video recorder $= \frac{22}{40}$. The percentage is

$$\frac{22}{40} \times 100\% = \frac{2200\%}{40} = 55\%$$

b Two shops are offering a sale on the price of the same type of television set. Which set is cheaper? Show your working.

'Notice you must give a mathematical explanation'

Solution

You will notice you are not given the actual price of the TV set. This does not matter. You can either change 30% to a fraction or decimal, or change $\frac{2}{5}$ to a percentage.

$$\frac{2}{5} \times 100\% = \frac{200\%}{5} = 40\%$$

Hence Dave is the cheaper – he gives the bigger reduction.

c A piece of wood 1.2 m long is cut into three pieces of length 60, 50 and 10 cm.

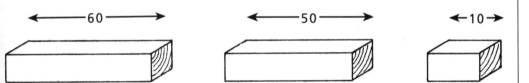

Express the length of the shortest piece as a percentage of the complete length. Give your answer correct to one decimal place.

Solution

First change 1.2 m to 120 cm. So you want to find 10 cm as a percentage of 120 cm. This is

$$\frac{10}{120} \times 100\% = \frac{1000\%}{120} = 8.3\%$$

d The price of a tin of beans was increased from 25p to 30p. What is this as a percentage?

Solution

The increase in price is

$$30p - 25p = 5p$$

The percentage increase is

$$\frac{5}{25} \times 100\% = \frac{500\%}{25} = 20\%$$

'The units of each quantity must be the same'

29

4 **a** In a class of 28, 10 are girls. Express as a ratio the number of boys to the number of girls, giving your answer as simply as possible.

The ratio must be in the correct order

Solution

The number of boys is $28 - 10 = 18$
So the ratio of boys to girls is $18 : 10$
which is the same as $9 : 5$ (by dividing each number by 2)

b Jean aged 10 and Paul aged 8 were left £450 between them by their uncle to be divided in the ratio of their ages. How much did they each receive?

Solution

The ratio of their ages is $10 : 8$
which is the same as $5 : 4$ (by dividing each number by 2)
The number of parts is $5 + 4 = 9$
One part is $£450 \div 9 = £50$
So Jean received $5 \times £50 = £250$
Paul received $4 \times £50 = £200$

Be sure to check that $£250 + £200 = £450$.

c In order to make the cricket club sandwiches for 12 people, the following ingredients are needed:

Eggs	3	Butter	6 oz
Loaves	3	Cheese	12 oz
Tomatoes	6	Lettuce	3

What ingredients are needed to make the same sandwiches for 16 people?

Solution

You do not have to find the amount for one person

To feed 16 people instead of 12 means that the quantities involved are in the ratio of $16 : 12$ which, when simplified, is $4 : 3$. This means that we need to multiply the original quantities by $\frac{4}{3}$. One way of doing this is to divide the amount of each ingredient by 3 and then multiply by 4.

Eggs	$3 \div 3 = 1$	Butter	$6\,oz \div 3 = 2\,oz$
	$1 \times 4 = 4$		$2\,oz \times 4 = 8\,oz$
Loaves	$3 \div 3 = 1$	Cheese	$12\,oz \div 3 = 4\,oz$
	$1 \times 4 = 4$		$4\,oz \times 4 = 16\,oz$
Tomatoes	$6 \div 3 = 2$	Lettuce	$3 \div 3 = 1$
	$2 \times 4 = 8$		$1 \times 4 = 4$

Do It Yourself

Answers are given in the Answers section at the back of the book.

1 a Shade the following shapes to agree with the fractions underneath:

2/3

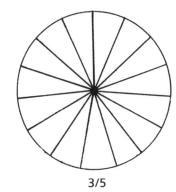

3/5

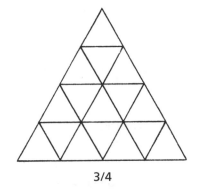

3/4

b When a car wheel with a puncture rests on the road, about $\frac{3}{20}$ of the circumference is flat. If the circumference of the wheel is 160 cm, what length is in contact with the ground?

c Change into mixed numbers:

(i) $\frac{7}{4}$ (ii) $\frac{15}{2}$ (iii) $\frac{19}{5}$

d Change the following mixed numbers into improper fractions:

(i) $2\frac{2}{3}$ (ii) $3\frac{1}{4}$ (iii) $5\frac{1}{4}$

2 a Change into a decimal:

(i) $\frac{4}{5}$ (ii) $\frac{3}{20}$ (iii) $\frac{3}{8}$

b Simplify the following fractions:

(i) $\frac{4}{6}$ (ii) $\frac{20}{35}$ (iii) $\frac{18}{24}$

c Find as a fraction in its simplest form:

(i) 0.6   (ii) 0.24   (iii) 0.65

d Change into a percentage:

(i) $\frac{4}{5}$ (ii) $\frac{3}{20}$ (iii) 0.24

3 a Ahmed asked his teacher how many marks he needed to pass the examination. His teacher told him the pass mark was 35%, and the total mark available was 80. How many marks does Ahmed have to get to pass?

b Find:
(i) 40p as a percentage of £2.00;
(ii) 75 cm as a percentage of 4 m;
(iii) £80 as a percentage of £2000.

c Two friends, Eleanor and Matthew, received pay rises. Eleanor's weekly wage increased from £120 to £138, while Matthew's weekly wage increased from £90 to £108. Who was given the greater percentage rise? Show all your working.

4 a Mortar is mixed from cement and sand in the ratio 5 : 2. How much sand would be needed to make a total of 84 kg of mortar?

b Sophie is making some cushions to sell at the Christmas Fayre. She needs 4.5 m of ribbon to make 6 cushions. How much will she need to make 15 cushions? The ribbon costs 80p for 1 m. How much will it cost her to buy the ribbon for the 15 cushions?

Estimating and Approximating

Things You Need to Know

1 How to **round** a number to a given degree of accuracy.

How to round a number to a given number of **decimal places** (count to the right of the decimal point).
How to round a number to a given number of **significant figures**.

2 How to estimate an answer to a calculation by rounding numbers to a suitable degree of accuracy. (This can be helpful in spotting mistakes you may have made when using a calculator.)

3 Solving problems using **trial and improvement**.

How to Do It

1 **a** Round the following numbers to one decimal place:
   (i) 8.634   (ii) 7.45   (iii) 8.07   (iv) 0.0362

Solution

(i) 8.6⎮34   6 is followed by 3 and so the 6 remains unaltered

∴ 8.634 = 8.6 to one decimal place

(ii) 7.4⎮5   4 is followed by 5 and so the 4 rounds up to 5

∴ 7.45 = 7.5 to one decimal place (abbreviate to 1 d.p.)
         = 7.5 (1 d.p.)

(iii) 8.0⎮7   0 is followed by 7 and so the 0 rounds up to 1

∴ 8.07 = 8.1 (1 d.p.)

(iv) 0.0⎮362   0 is followed by 3 and so the 0 remains unaltered

∴ 0.0362 = 0.0 (1 d.p.)

b Round the following numbers to two significant figures:
   (i) 493   (ii) 0.04873   (iii) 4.093

Solution
Remember that significant figures are always counted from the first digit
that is not a zero.

(i) 49⎮3   9 is followed by 3 and so remains unaltered

∴ 493 = 490 to two significant figures (abbreviate to 2 s.f.)

(ii) 0.048⎮73   8 is followed by 7 and so rounds up to 9

∴ 0.04873 = 0.049 (2 s.f.)

(iii) 4.0⎮93   0 is followed by 9 and so rounds up to 1

∴ 4.093 = 4.1 (2 s.f.)

'Significant figures are the figures in any number which give you extra information'

c How many 23p stamps can you buy for £2?

Solution
Change £2 into 200p. Now

$$200 \div 23 = 8.696$$

So 8 stamps can be bought.

'Here we have rounded down'

2a Jamie and Elaine ran a weekend job cleaning cars. They charged £2.80 for a complete clean. They estimated that in a year they would clean 6 cars on about 36 weekends. Roughly how much will they earn in one year?

Solution
The exact total from their figures would be $36 \times 6 \times £2.80$. Since we cannot use a calculator, then we can estimate the answer by making the numbers easier. So the estimated total is

$$40 \times 5 \times £3 = 200 \times £3 = £600$$

(The exact answer is £604.80. A pretty good estimate!)

b Jean was showing off to her friends that she could multiply any pair of two digit numbers in her head in less than 10 seconds. They asked her to work out 98×47. She gave the answer 4604. Explain how they could tell her quickly that this was not the exact answer.

Solution
You can often see a mistake by looking at the *last* digits in a calculation.
    Since $8 \times 7 = 56$, the answer must end in 6. Her answer ends in 4 and so must be wrong.

3a Use a trial and improvement method to solve the equation $x^3 = 16$, to two decimal places.

Solution

Try $x = 1$	$1^3 = 1$	which is a lot less than 16
Try $x = 2$	$2^3 = 8$	which is still less than 16
Try $x = 3$	$3^3 = 27$	which is more than 16

'This is the same as finding $\sqrt[3]{16}$'

35

Hence x lies between 2 and 3.

Try $x = 2.5$ $2.5^3 = 15.625$, which is nearly 16
Try $x = 2.51$ $2.51^3 = 15.813$
Try $x = 2.52$ $2.52^3 = 16.003$

So x is between 2.51 and 2.52, and much closer to 2.52 than to 2.51.
Hence $x = 2.52$ rounded to two decimal places.

b David was solving the equation $x^3 + x = 40$. His teacher told him that there was a solution between $x = 3$ and $x = 5$. Use a trial and improvement method to find this solution correct to 1 decimal place.

Solution
Look at the teacher's values first:

$x = 3$ $3^3 + 3 = 30$ which is too low
$x = 5$ $5^3 + 5 = 130$ which is much too high
Try $x = 3.5$ $3.5^3 + 3.5 = 46.375$ which is still too high
Try $x = 3.4$ $3.4^3 + 3.4 = 42.704$ which is too high
Try $x = 3.3$ $3.3^3 + 3.3 = 39.237$ which is just too low

The nearest answer is $x = 3.3$

Do It Yourself

Answers are given in the Answers section at the back of the book.

1

a Round to the nearest whole number:

(i) 8.6   (ii) 28.3   (iii) 0.46

b Round to the nearest hundred:

(i) 863   (ii) 950   (iii) 9.6

c Round to the nearest centimetre:

(i) 6.9 cm   (ii) 28 mm   (iii) 3.45 cm

d Round the following numbers to one decimal place:

(i) 6.43   (ii) 8.06   (iii) 0.995

e Round to two significant figures:

    (i) 864   (ii) 6.093   (iii) 38 600

f A small glass holds about 65 ml of liquid. A bottle which holds 1.5 litres of a fruit drink is emptied into a number of these glasses. How many can be filled? (1 litre = 1000 ml)

g The diagram shows a leaf that Susan is measuring in a science experiment. What is the length of the leaf (including the stalk) rounded
   (i) to the nearest centimetre;
   (ii) to the nearest half-centimetre;
   (iii) to the nearest millimetre.

h A crowd of 38 434 attended the match between Manchester United and Arsenal. Round this attendance:
   (i) to the nearest 500;
   (ii) to the nearest 1000;
   (iii) to the nearest 10 000.

2 a Use sensible rounding to estimate the value of the following:

    (i) $8.9 \div 3.17$ (ii) $865 \div 296$
    (iii) $98.4 - 37.6$ (iv) 0.086×193

b Use the last digit in the following calculations to pick out those which *must* be wrong:

    (i) $562 \times 18 = 10\,116$ (ii) $87 \times 97 = 8434$
    (iii) $68 \times 77 = 5036$ (iv) $63 \times 89 = 5607$
    (v) $253 \times 45 = 11\,358$

3 a Use a trial and improvement method to solve the equation $x^3 = 23$ correct to three significant figures.

b Solve the equation $x^4 + x^2 = 10$ (giving two decimal places in your answer), and find the solution that lies between $x = 1$ and $x = 2$.

5 Units and Conversion Graphs

'A measurement like 3 ft 5 in is sometimes written as 3' 5"'

Things You Need to Know

1 **The metric system**:
  (i) units of length: millimetres (mm), centimetres (cm), metres (m) and kilometres (km):

  1 cm = 10 mm   1 m = 100 cm   1 m = 1000 mm   1 km = 1000 m

  (ii) Units of weight: grams (g), kilograms (kg), tonnes:

  1 kg = 1000 g   1000 kg = 1 tonne

  (iii) Units used to measure capacity (usually volume of liquids): millilitre (ml) or sometimes cubic centimetre (cm³), the litre (l) and the centilitre (cl):

  1 l = 1000 ml or 1 l = 100 cl   1 cl = 10 ml

2 **The imperial system**:
  (i) Units of length:

  12 inches (in) = 1 foot (ft)
  1 yard (yd)   = 3 feet (ft)
  1 mile   = 1760 yards (yds)

(ii) Units of weight:

> 16 ounces (oz) = 1 pound (lb)
> 14 pounds   = 1 stone
> 8 stone   = 1 hundredweight (cwt)
> 1 ton   = 20 cwt

(iii) Units of capacity:

> 1 gallon = 8 pints
> 1 fluid oz of water weighs 1 oz

3 There are some approximate conversions between metric and imperial units that are quite useful:

> 1 lb = 450 g   20 fluid oz = 1 pint
> 1 kg = 2.2 lb   1 gallon   = 4.5 l
> 1 ton = 1 tonne

'You will usually be told this information'

4 Different units can be converted by means of a **conversion graph**.

5 How to use compound units such as km/h and miles per gallon.

How to Do It

1 **a** (i) Convert 240 mm into metres.
     (ii) Convert 8.43 kg into grams.
     (iii) Convert 2.4 l into centilitres.

Solution

(i) Since 1 m = 1000 mm, you must divide by 1000. So

> 240 ÷ 1000 = 0.24 m

(ii) Since 1 kg = 1000 g, you must multiply by 1000. Hence

> 8.43 × 1000 = 8430 g

(iii) Since 1 l = 100 cl, you must multiply by 100. So

> 2.4 × 100 = 240 cl

'As a first step, try to decide whether the answer should be a larger or smaller number'

'When you want to measure a liquid, you measure the amount of space the liquid takes up. This is called its capacity'

b A medicine bottle holds 1.25 l. A chemist takes 25 ml doses from this bottle using a syringe. How many doses can the chemist obtain from the bottle?

Solution
You must first change the volume of the bottle into litres, using the conversion: litres → millilitres is × 1000.

So the bottle holds $1.25 \times 1000 = 1250$ ml. The number of doses is

$$1250 \div 25 = 50 \text{ doses}$$

2 How many inches are there in 2 yds 1 ft 4 in?

Solution

$$2 \text{ yds} = 2 \times 3 = 6 \text{ ft}$$

So the distance is 7 ft 4 in.

$$7 \text{ ft} = 7 \times 12 = 84 \text{ in}$$

The total distance $= 84 + 4 = 88$ in.

3 Use the facts that 1 in = 2.54 cm, and 1 mile = 1760 yds, to find the number of metres in 1 mile.

Solution
This is not an easy question, and a careful plan is needed. We know:

$$1 \text{ yd} = 3 \text{ ft}$$
$$1 \text{ ft} = 12 \text{ in}$$

Hence $1 \text{ yd} = 3 \times 12 = 36 \text{ in}$.

$$\therefore 1 \text{ yd} = 36 \times 2.54 = 91.44 \text{ cm}$$

'There is no point in giving 1609.344 as the answer'

So $1 \text{ mile} = 1760 \times 91.44 = 160\,934.4$ cm.

To change centimetres to metres, you divide by 100. Hence

$$1 \text{ mile} = 160\,934.4 \div 100 \text{ m}$$
$$= 1609 \text{ m (to the nearest metre)}$$

4 The graph at the top of page 41 shows the conversion between £ and French francs (fr.)

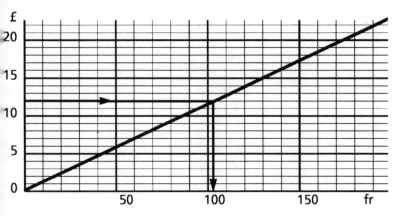

a How many francs will you get for: (i) £12   (ii) £35

b How many £ will you get for: (i) 60 fr.   (ii) 350 fr.

Solution

a (i) This is shown on the graph. Read *across* horizontally from £12, and then from the conversion line *down* vertically to get 103 fr.

(ii) £35 does not appear on the graph, and must be obtained in separate parts. So

$$£20 \text{ is } 170 \text{ fr.}$$
$$£15 \text{ is } 127 \text{ fr.}$$

Total   £35 is 297 fr.

b (i) Read *up* vertically from 60 fr. to the conversion line, and *across* horizontally to get £7.

(ii) Here also 350 fr. does not appear on the scale. But 50 fr. = £6, so

$$350 \text{ fr.} = 7 \times 50 \text{ fr.} = 7 \times £6 = £42$$

5 a A car travels 16 km in 12 minutes. Find the average speed of the car in km/h.

Solution
Method 1
This method works well if the time taken divides exactly into 60 minutes.

Since $60 \div 12 = 5$, then 12 minutes is $\frac{1}{5}$ of an hour. Hence in 1 hour, the car travels $16 \times 5 = 80$ km. So the average speed $= 80$ km/h.

'Remember km/h means how many kilometres are travelled in one hour'

41

Method 2

Since

$$\text{Average speed} = \frac{\text{distance travelled}}{\text{time taken}}$$

$$= \frac{16}{12}\,\text{km/min}$$

To change this into km/h, multiply by 60. Hence

$$\text{Average speed} = \frac{16}{\cancel{12}_1} \times \cancel{60}^{5} = 80\,\text{km/h}$$

b Alec cycles for 24 minutes, at an average speed of 18 km/h. Calculate the distance he has travelled.

Solution

You need to be careful here because the units are not consistent. It is best to change 24 minutes into $\frac{24}{60}$ hours. Using distance = speed × time, we have

$$\text{Distance} = 18 \times \frac{24}{60} = 7.2\,\text{km}$$

Do It Yourself

Answers are given in the Answers section at the back of the book.

1 Complete the following:

    (i) 464 cm = ... m   (ii) 63 mm = ... cm

    (iii) 2.85 m = ... cm   (iv) 0.6 m = ... mm

    (v) 6200 g = ... kg   (vi) 1.25 l = ... ml

    (vii) 260 ml = ... l   (viii) 8.9 kg = ... g

2 Complete the following:

    (i) 2 ft 3 in = ... in   (ii) $4\frac{1}{2}$ yds = ... ft

    (iii) $3\frac{1}{2}$ gallons = ... pints   (iv) $2\frac{1}{2}$ stone = ... lb

    (v) 1 stone = ... oz   (vi) 100 oz = ... lb

    (vii) 2 pints = ... gallons   (viii) 88 in = ... ft ... in

3a Use the fact that 1 cm = 0.39 in to convert:
   (i) 8 cm into inches   (ii) 2 m into inches

b Paul and Mark are trying to convert the fuel consumption of their dad's car from 30 miles/gallon to kilometres/litre.

   (i) Use the fact that 1 mile = 1.6 kilometres to find how many kilometres the car travels on 1 gallon of petrol.
   (ii) Use the fact that 1 gallon = 4.5 litres to convert the fuel consumption into kilometres/litre.

4a Construct a conversion graph between pounds and kilograms, using the fact that 1 kg is approximately 2.2 lb. (Use a scale of 0–20 kg, and 0–45 lb.) Use your graph to make the following conversions:

   (i) 8.6 lb to kg;   (ii) 50 lb to kg;   (iii) 15 kg to lb;   (iv) 85 kg to lb.

'Use graph or squared paper here'

b The graph shown will convert miles/hour into metres/second. Use the graph to answer the following questions:

   (i) Convert 40 miles/hour into metres/second.
   (ii) Convert 28 metres/second into miles/hour.
   (iii) How many hours would it take to travel 100 miles travelling at 20 metres/second?

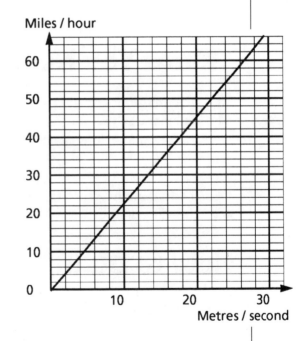

5a A train runs between two stations at an average speed of 87 km/h. If the journey time is 1 hour 20 minutes, what is the distance between stations?

b A coach service runs between two towns A and B on a motorway 180 km apart. The coach leaves A at 0824, and arrives at B at 1054. Calculate the average speed of the coach on its journey between A and B.

6 | Measuring

Things You Need to Know

1 The units that you choose are important because the correct choice avoids too many zeros after the decimal point (e.g. 0.0004 km) or too many zeros in front of the decimal point (e.g. 20 000 mm).

2 If information is given to you in a variety of units, change all of them into the same units before making any calculations.

How to Do It

1 **a** Which units would you use to measure the following quantities:
   (i) The measurements of a garden
   (ii) The amount of fat in a cake
   (iii) The volume of a wine glass
   (iv) Your height
   (v) Reaction times

Solution

(i) The **metre** would be the most sensible, although some people would use feet or yards.

(ii) Since there would not be as much as 1 kg in a cake, the best unit to use would be the **gram**.

(iii) The litre would obviously be too great, hence you would probably choose the **millilitre**. However, the **centilitre**, which is 10 ml, is probably even better.

(iv) Although your height is not an exact number of metres, the **metre** is the best unit.

(v) This will be a small time interval. The best unit would be the **second**.

b Change the following measurements into more sensible ones:

     (i) 20 000 mm   (ii) 0.086 kg   (iii) 244 in

     (iv) 0.00263 m   (v) 1880 minutes   (vi) 100 000 pence

Solution

(i) Since 1 m = 1000 mm, if we divide by 1000, we get

$$20\,000 \div 1000 = 20\,m$$

(ii) Since 1 kg = 1000 g, if we multiply by 1000, we get

$$0.086 \times 1000 = 86\,g$$

(iii) Since 12 in = 1 ft, $244 \div 12 = 20$ remainder 4. So 244 in = 20 ft 4 in.

     Since 3 ft = 1 yd and $20 \div 3 = 6$ remainder 2, the best answer would be 6 yds 2 ft 4 in.

(iv) This needs to be made larger, so change it into millilitres, i.e. multiply by 1000:

$$0.002\,63 \times 1000 = 2.63\,mm$$

(v) 60 minutes = 1 hour

     $1880 \div 60 = 31$ remainder 20

     24 hours = 1 day

     So 1880 minutes = 1 day 7 hours 20 minutes

(vi) £1 = 100 pence

     $100\,000 \div 100 = 1000$

     100 000 pence = £1000

'Try to measure items yourself to get the feel of how large these units are'

2 a What is the total length of the bolt shown in the diagram?

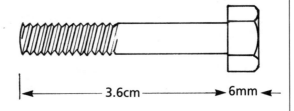

3.6cm   6mm

*'Always
change the
units that are
easiest to
change'*

Solution

The units are not the same, and so you must alter one of them. So

$$3.6\,\text{cm} = 36\,\text{mm}$$
$$\text{Length} = 36 + 6 = 42\,\text{mm}$$

You could also give the answer as 4.2 cm.

b Tiles measuring 80 cm by 60 cm are
being laid on an area that measures
4 m by 3 m. How many tiles will be
needed?

Solution

First change the measurements of
the area into centimetres. This is 400
by 300. The tiles must be laid the
correct way round, or else they will
not fit exactly. They must be laid as
in the diagram. Now

$$400 \div 80 = 5 \quad \text{and} \quad 300 \div 60 = 5$$

Hence there will be 5 rows of 5 tiles.
The total $= 5 \times 5 = 25$ tiles.

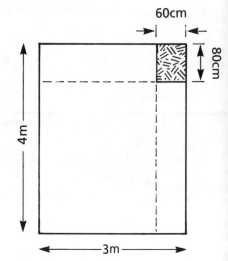

Do It Yourself

Answers are given in the Answers section at the back of the book.

1 a Which units would you use to measure the following quantities:

  (i) The volume of a bucket
  (ii) The thickness of a sheet of glass
  (iii) The weight of a car
  (iv) The length of a railway coach
  (v) The length of the Channel tunnel
  (vi) Cooking times

b Change the following measurements into more sensible ones:

  (i) 400 mm   (ii) 0.0892 kg   (iii) 2000 yds
  (iv) 0.16 cm   (v) 10 000 cl   (vi) 2 000 000 g
  (vii) 0.8 hours

2 a The diagram shows a target at a shooting range. The diameter is 1.8 m, and the widths of the outside regions are 40 cm and 30 cm as shown. What is the radius of the centre bull?

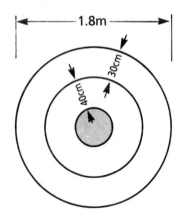

b A box measures 1.2 m by 0.7 m by 60 cm. It is filled with wooden blocks that measure 14 cm by 12 cm by 5 cm. How many blocks can be put in the box?

7 Statistics

Things You Need to Know

1 How to design a **questionnaire** to survey opinion (taking account of bias).

2 How to collate your results in a **tally chart**.

3 How to present your results, using **pie charts**, **bar charts** and **line graphs**.

4 How to construct a **scatter diagram**, and comment on the type of **correlation** shown.

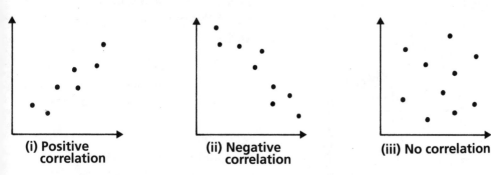

(i) Positive correlation

(ii) Negative correlation

(iii) No correlation

How to Do It

1 Class 9 are devising a questionnaire to survey the likes and dislikes of types of sweets eaten by children in their school.

From the list of proposed questions given below, comment on which questions are good and which are poor.

> (i) Do you like chocolate?
> (ii) Does your mum buy lots of sweets?
> (iii) Do you prefer wrapped or unwrapped sweets?
> (iv) Don't you think sweets are expensive?
> (v) How many times a week do you eat sweets?

Solution

(i) A good question, needs only yes/no.

(ii) Can be answered yes/no, but probably has nothing to do with the survey.

(iii) A poor question. It is unlikely you could answer this question exactly.

(iv) A poor question. It pushes a person to say 'yes'.

(v) This question needs to be accompanied by a choice of numbers, and also the phrase 'on average' should be inserted.

Class	Tally	Frequency
21–25	I	1
26–30	IIII	4
31–35	II	2
36–40	IHT III	8
41–45	IHT I	6
46–50	III	3
		Total 24

2 The marks out of 50 scored by a group of students in a maths test are given below. Use a tally chart to record this data using suitable class widths.

40	25	31	46	38	39	45	36
42	30	37	43	40	29	47	50
35	37	43	29	26	41	38	43

Solution
The lowest mark is 25, the highest 50. To make sure the classes are the *same size* they need to be as shown in the table on the left.

'Make sure the total is correct'

3 Use the data in question 2 and present them using (i) a pie chart and (ii) a bar chart.

Solution
(i) Method 1
Divide 360 by the total 24: $360 \div 24 = 15$

Hence each pupil is represented by 15°.
The angles for each sector are as follows:

Frequency	1	4	2	8	6	3
Angle	15°	60°	30°	120°	90°	45°

'Use this method if the total divides exactly into 360'

Method 2
To find the angle for each sector, you work out

$$\frac{\text{frequency}}{\text{total}} \times 360$$

'Make sure the pie chart is labelled'

Hence for the 36–40 class, the angle would be

$$\frac{8}{24} \times 360^{15} = 120°$$

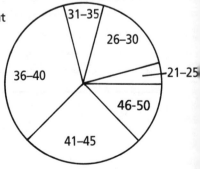

In either case the pie chart would be as shown here.

(ii)

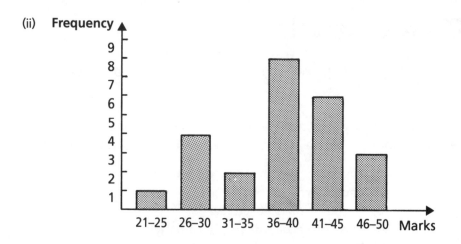

'The bars do not have to touch each other in a bar chart'

4 A group of 10 children took tests in mathematics and physics. The scores obtained are given in the following table:

Maths score	25	40	36	45	38	12	18	46	30	38
Physics score	20	38	38	40	27	15	20	48	31	36

(i) Illustrate these results on a scatter diagram.
(ii) Describe the correlation shown in the scatter diagram.

Solution

(i)

The results are plotted with crosses on a clearly labelled diagram.

(ii) The diagram shows positive correlation.

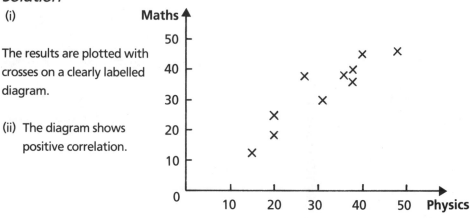

Do It Yourself

Answers are given in the Answers section at the back of the book.

1 Say which of the following questions in a questionnaire would be good, and which would be poor. Give a reason for your answer in each case.

    (i) You must agree that smoking is bad for you?
    (ii) How often do you play tennis?
    (iii) Do you have a video recorder?
    (iv) How many calories do you eat each day?
    (v) Do you usually eat more than 1500 calories each day?
    (vi) Do you collect stamps?
    (vii) What sort of music do you like?
    (viii) What is your favourite subject at school?

2 Rosie is carrying out a survey on the length (in cm) of pea pods. She noted her results on a sheet of paper as follows:

```
5.6   6.3 | 4.1   5.1
4.8 4.9 | 4.9 3.3
   2.8   6.6
6.8  5.4 |      5.9
                4.7
3.9   6.1 4.7  3.9
   5.1    4.6
```

Class	Tally	Frequency
$2 \leqslant x < 3$		
$3 \leqslant x < 4$		
$4 \leqslant x < 5$		
$5 \leqslant x < 6$		
$6 \leqslant x < 7$		

Complete the tally chart shown alongside for the length x of the pods.

3 Using the results of question 2:

(i) Represent data in both a pie chart and a bar chart.

(ii) Say which of the diagrams you think is better. Give a reason.

4 The number of ice creams sold by Marco's in a day, and the midday temperature are given in the following table:

No. of ice creams	84	120	110	102	100	140	110	95
Temperature (°C)	20	24	27	19	22	26	20	21

(i) Plot these data on a suitable scatter diagram.
(ii) Describe the correlation shown.
(iii) Would this agree with what you would expect?
(iv) Give a possible explanation for your answer to part (iii).

'It's worth checking that all the segments add up to 360° now before you draw it out'

8 Probability

Things You Need to Know

1 **Probability** measures how likely it is that something will happen. Probability is given as a fraction or decimal on a scale of 0 to 1.

$$\text{Probability} = \frac{\text{number of favourable outcomes}}{\text{total number of outcomes}}$$

Probabilities can be found from (i) theory; (ii) experiment; (iii) survey; or (iv) looking at past data.

Random events are equally likely to happen. When you are carrying out a survey, if your data are not from a random sample, then **bias** will be introduced, which will distort the results.

2 Mathematicians use the word 'expect' in a special way. They really mean what you would get if you repeated the experiment and averaged the results over a long period of time. The **expected** value of an experiment in probability can be found by using the result:

Expected number of outcomes = probability of that outcome
× total number of trials of the experiment

3 The total of all possible probabilities in a given situation is 1. The probability of something not happening = 1 − the probability that it does happen.

4 How to find the probability of an outcome combined from two independent outcomes. For example, the total score if you roll two dice.

How to Do It

1 There are different ways of estimating probabilities.
Method A: Use equally likely outcomes
Method B: Look back at data
Method C: Carry out a survey or experiment

State which method is required to estimate the following probabilities:
  (i) The probability that a pupil in your school has a pet rabbit.
 (ii) The probability that I will win a raffle if I buy 5 tickets and I know that 1000 are sold.
(iii) The probability that it will rain tomorrow.
(iv) The probability that a drawing pin will land point up if dropped on to a table.

Solution
  (i) It is obviously not sensible to ask everybody, and so you would use method C.
 (ii) You know the exact numbers involved, and hence method A is used.
(iii) This would be based on past experience. Use method B.
(iv) You could only do this by experiment, and so method C is appropriate.

'Each ticket is equally likely to win'

2ª A six-sided die is rolled 180 times. Approximately how many times would you expect the number showing on the top to be a 5?

*'You may get
an answer
quite different
if you try this'*

Solution

The probability of scoring $5 = \frac{1}{6}$. Hence the expected number of 5s is

$$\frac{1}{6} \times 180 = 30$$

b In order to estimate the number of pupils likely to come from outside the village next year, a school looked back at its records over the last few years. They found that 250 lived in the village, and 550 lived outside. If 160 children are likely to attend the school next year, approximately how many will live in the village?

Solution

In the data, the total number of children is $250 + 550 = 800$. The probability of living in the village is

$$\frac{250}{800} = \frac{5}{16}$$

Hence next year, the expected number of children living in the village is

$$\frac{5}{16} \times 160 = 50 \text{ children}$$

*'You are
assuming next
year is similar
to the past'*

3a When Richard and Katie play each other at chess, the probability that Richard wins is $\frac{1}{4}$ and the probability that Katie wins is $\frac{1}{3}$. What is the probability that they will draw?

Solution

There are no possibilities other than win, lose or draw, hence the three probabilities must add up to 1. Now

$$\frac{1}{4} + \frac{1}{3} = \frac{3}{12} + \frac{4}{12} = \frac{7}{12}$$

So the probability of a draw is

$$1 - \frac{7}{12} = \frac{5}{12}$$

b 15 out of 50 people have blue eyes. What is the probability that a person chosen at random does not have blue eyes?

Solution
Method 1
The probability that the person does have blue eyes is

$$\frac{15}{50} = \frac{3}{10}$$

The probability that the person does not have blue eyes is

$$1 - \frac{3}{10} = \frac{7}{10}$$

Method 2
If 15 people do have blue eyes, then $50 - 15 = 35$ people do *not* have blue eyes. The probability of not having blue eyes is

$$\frac{35}{50} = \frac{7}{10}$$

'Choose whichever method you prefer'

4 A game is played using two dice (red and blue) in the shape of a pyramid. Each of the four faces is numbered 1 to 4. The two dice are thrown and your score is obtained by multiplying together the numbers that land face down. Hence if 4 and 2 lie face down, your score is $4 \times 2 = 8$. List all the ways in which the dice can land and the scores you get. What is the probability of getting a score less than 10?

Solution
Try to be systematic about listing the outcomes:

(1, 1) $1 \times 1 = 1$	(1, 2) $1 \times 2 = 2$	(1, 3) $1 \times 3 = 3$	(1, 4) $1 \times 4 = 4$
(2, 1) $2 \times 1 = 2$	(2, 2) $2 \times 2 = 4$	(2, 3) $2 \times 3 = 6$	(2, 4) $2 \times 4 = 8$
(3, 1) $3 \times 1 = 3$	(3, 2) $3 \times 2 = 6$	(3, 3) $3 \times 3 = 9$	(3, 4) $3 \times 4 = 12$
(4, 1) $4 \times 1 = 4$	(4, 2) $4 \times 2 = 8$	(4, 3) $4 \times 3 = 12$	(4, 4) $4 \times 4 = 16$

There are 16 possible outcomes, and 13 of them give a score less than 10. The probability of scoring less than 10 is $\frac{13}{16}$.

'The score on each die is independent of the other'

Do It Yourself

Answers are given in the Answers section at the back of the book.

1 **a** Find the probability of the following events happening:

   (i) An ace being drawn at random from a complete set of playing cards.
   (ii) Two coins both landing heads if you spin them.
   (iii) The probability that someone you spoke to at random had the same birthday as yours.
   (iv) The probability that a die shows 2 if rolled.
   (v) The probability that a two-digit number written down at random ends in 3.
   (vi) The probability that a coin lands heads, if it has just landed tails eight times consecutively.

b Ibrahim was carrying out a car survey along the road outside the school. He recorded the colours of the cars in a table as follows:

Colour	Red	White	Blue	Green
Number	4	7	8	6

From his results, what is the probability that a car randomly selected on the road is white?

2 **a** Using the results from question 1(b), if Ibrahim counted 200 cars during the morning, how many of them are likely to be blue?

b A game at a fairground is marked with five equal sectors as shown in the diagram.

   (i) What is the probability of winning?
   (ii) If you had 10 goes, how many times would you get your money back?

3 a The local hockey team has won five of the last nine matches. What is the probability that it will not win the next match?

b A box of sweets contains chocolates, jellies and fruit gums. The probability of getting a jelly is $\frac{1}{5}$ and the probability of getting a chocolate is $\frac{1}{4}$. What is the probability of getting a fruit gum?

4 Two ordinary dice are rolled. Draw a suitable diagram to show all the possible outcomes. The score is obtained by adding the two numbers showing on top of the dice. Find:

(i) the probability of scoring 6;

(ii) the probability that your score is a multiple of 3.

9 | Angles and Triangles

Things You Need to Know

1 Angles:
   (i) There are 360 degrees (°) in a complete turn.
   (ii) Angles less than 90° are called **acute**.
   (iii) Angles between 90° and 180° are called **obtuse**.
   (iv) Angles greater than 180° are called **reflex**.
   (v) An angle of 90° is called a **right angle**: symbol ∟.
        Lines at right angles are **perpendicular**.
   (vi) Angles on a straight line add up to 180°.

$a + b = 180°$

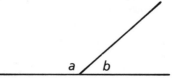

   (vii) Angles at a point add up to 360°.

$a + b + c + d = 360°$

   (viii) The angles of a triangle add up to 180°.

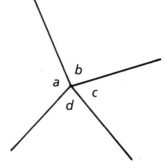

2 Parallel lines (marked with arrows)

$a = b$: **opposite angles**

$\left.\begin{array}{l} a = c \\ d = e \end{array}\right\}$ **alternate angles** (sometimes called 'z' angles)

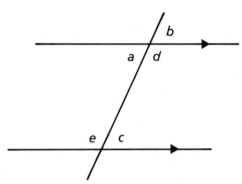

$c = b$: **corresponding angles** (sometimes called 'f' angles)

$\left.\begin{array}{l} d + c = 180° \\ a + e = 180° \end{array}\right\}$ **interior angles**

3 How to construct a triangle accurately using a ruler, pair of compasses and a protractor. (Make sure you have a sharp pencil, and that you can clearly read the measurements on the ruler and protractor.)

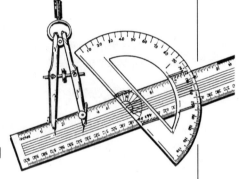

Types of triangles:
  (i) **Scalene**: all sides and angles different
 (ii) **Isosceles**: two equal sides and two equal angles
(iii) **Equilateral**: all sides equal and all angles equal

4 How to represent a solid in two dimensions, by drawing a plan or elevation, or by using isometric paper.

How to Do It

1 In the diagram, find the value of x.

'Notice an equation is needed'

Solution

Since angles at a point add up to $360°$, then

$$x + 2x + 110° + 100° = 360°$$

Simplifying this,

$$3x + 210° = 360°$$

So $\qquad 3x = 360° - 210° = 150°$

Hence $\qquad x = 50°$

2a Find angles a, b, c in the diagram. Give reasons for your answers.

Solution

$a = 180° - 80° = 100°$ (angles on a straight line)

$a = b = 100°$ $\qquad$ (corresponding angles)

$c = b = 100°$ $\qquad$ (opposite angles)

b This question refers to the diagram alongside. Find angles x and y, giving reasons for your answer.

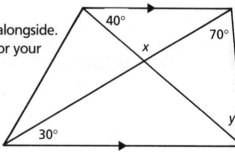

Solution

Some extra angles have been labelled to help.

$$p = 30° \quad \text{(alternate angles)}$$

Now $40° + p + x = 180°$ (angle sum of triangle)

$$\therefore \quad 40° + 30° + x = 180°$$

Hence $\qquad\qquad x = 110°$

'Make sure you state the reasons clearly'

Now $x + q = 180°$ (angles on a straight line)

$$\therefore \quad 110° + q = 180°$$

Hence $\quad\quad\quad q = 70°$

But $q + y + 70° = 180°$ (angle sum of triangle).
So

$$70° + y + 70° = 180°$$
$$\therefore \quad y = 40°$$

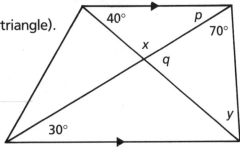

3a The diagram shows part of a bridge structure.
Make an accurate scale drawing using a scale of
1 cm to 1 m, and find the length of the parts
BD and DC.

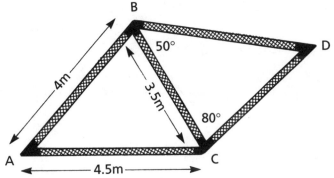

Solution

(i) Construct triangle ABC:

(a) Put compass point on A, radius 4 cm and draw an
   arc.
(b) Put compass point on C, radius 3.5 cm and draw
   an arc.

The two arcs meet at B.

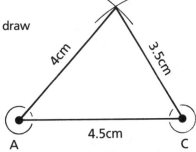

'You will only need one diagram'

(ii)

(a) Put the protractor at B, and draw a line at 50° as shown.
(b) Put the protractor at C, and draw a line at 80° as shown.
The lines will meet at D.

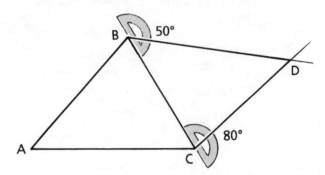

(iii)

By measuring, BD = 6.8 m
By measuring, DC = 3.5 m

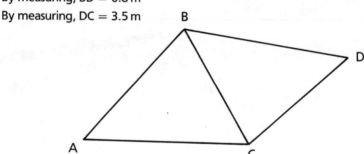

b State what type of triangle ABC is in the following cases:
  (i) angle A = angle B = angle C
  (ii) angle A = 70° and angle B = 40 °
  (iii) AB = 5 cm, BC = 6 cm, and AC = 8 cm.

Solution

  (i) Because all angles are the same, all sides will also be the same.
       Hence ABC is **Equilateral**.
  (ii) Since the angles of a triangle add up to 180°,

$$C = 180° - 70° - 40° = 70°$$

  Hence two angles are equal.
  The triangle ABC is **Isosceles**.
  (iii) Since all sides are of a different length, all angles will be different.
        Hence, triangle ABC is **Scalene**.

4 The diagrams show the front and side elevation of an iron casting.

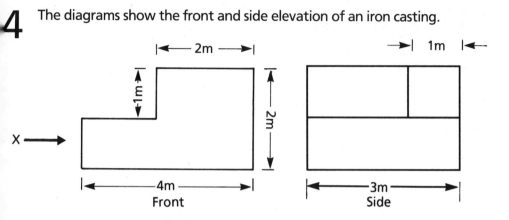

(i) Draw on isometric paper a possible solid.
(ii) Draw a possible plan of the object.

Solution

(i) A possible shape has been drawn on the isometric paper. The side elevation does not tell you how much has been cut out on the side you cannot see.

(ii) Think of a plan as a bird's eye view (from above). A possible plan is shown below.

'Isometric paper is extremely useful for three-dimensional shapes'

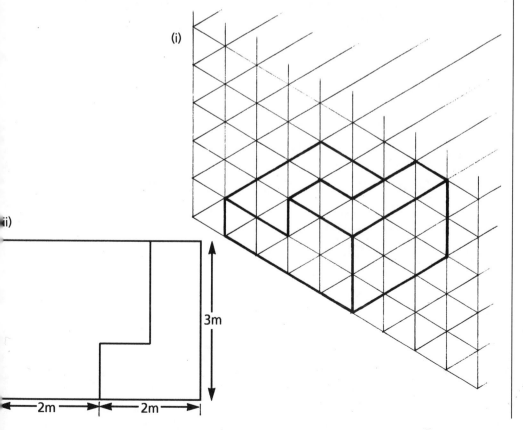

Do It Yourself

Answers are given in the Answers section at the back of the book.

1 Find *x* in the following diagrams:

(i)

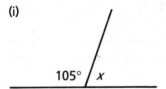

(ii)

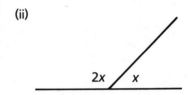

(iii)

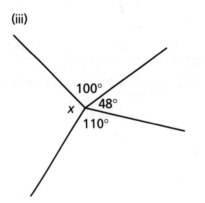

(iv)

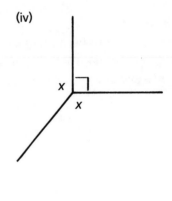

(v)

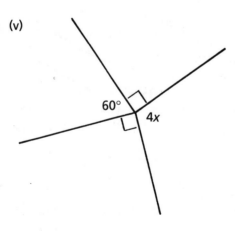

Find the marked angles in the following diagrams. Give a reason for your answer in each case.

(i)

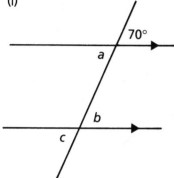

(ii)

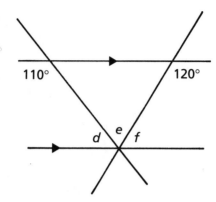

(iii)

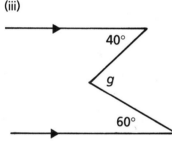

(iv)

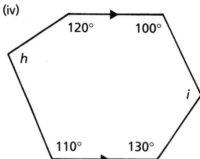

3 **a** Construct accurately triangle ABC in the following cases:

    (i) AB = 6 cm, angle A = 40°, angle B = 36°, measure BC.
    (ii) AB = 8.5 cm, AC = 6.8 cm, BC = 4.3 cm, measure angle A.
    (iii) AB = 8 cm, angle A = 60°, BC = 9.8 cm, measure angle B.

b State what type of triangle PQR is in the following cases:

  (i) angle P = 72°, angle Q is twice angle R
  (ii) PQ = QR = PR = 10 cm
  (iii) angle P = 63°, angle Q = 36°

4 Draw a front and side elevation of the solid shown below.

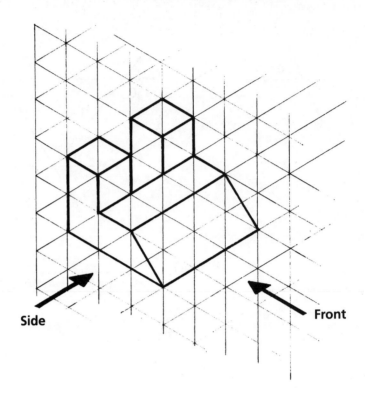

Side

Front

Coordinates, Maps and Bearings

Things You Need to Know

1 How to locate a point using **cartesian** coordinates (x, y).
Points plotted often show a **relationship** between the x and y coordinates. This relationship can be expressed using an equation or a mapping statement. For example,

$$y = 3x + 1$$
$$x \rightarrow x^2 + 1$$

2 Position can also be located using a three-figure **bearing**. A bearing is always measured from the north in a *clockwise* direction.

3 If a map is drawn it requires a **scale** to be stated, e.g. 1 : 20 000 means 1 cm on the map represents a real length of 20 000 cm.

How to Do It

1 Complete the table for the mapping

$$x \rightarrow 5 - 2x$$

Plot the values from the table on to a graph.

'*This means x = 0, y = 5, and so on*'

x	5 − 2x
0	
1	
2	1
3	
4	

Solution

If $x = 0$, $x \rightarrow 5 - 2 \times 0 = 5$
If $x = 1$, $x \rightarrow 5 - 2 \times 1 = 3$
If $x = 3$, $x \rightarrow 5 - 2 \times 3 = -1$
If $x = 4$, $x \rightarrow 5 - 2 \times 4 = -3$

'*The coordinates of a point are its address, if you like*'

The points $(0, 5)$, $(1, 3)$, $(2, 1)$, $(3, -1)$ and $(4, -3)$ can now be plotted on a graph as shown alongside.

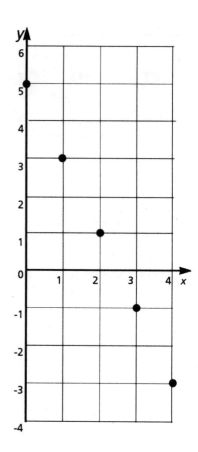

2 B is due south of a church at C, and A is due west of B. Calculate the bearing of A from the church.

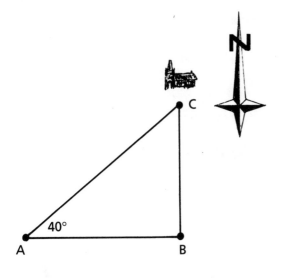

Solution

Remember that a bearing is measured from north in a clockwise direction.

Draw the north line at C; angle ACB = 50°, hence the bearing is

$$180° + 50° = 230°$$

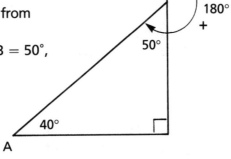

'Never do a question on bearings without a diagram'

3 A map is drawn to a scale of 1 : 25 000. Find:
 (i) the real length of a tunnel which is 4 cm long on the map;
 (ii) the length of a lake on the map which is 3 km long.

Solution

1 : 25 000 means 1 cm represents 25 000 cm but 1 km = 100 000 cm. So 1 cm represents

$$25\,000 \div 100\,000 = 0.25\,\text{km}$$

 (i) 4 cm will be $4 \times 0.25 = 1\,\text{km}$
 (ii) If 1 cm represents 0.25 km, then 4 cm represents 1 km. Hence a length of 3 km will be $4 \times 3 = 12\,\text{cm}$.

'1:25 000 is much easier to use if simplified to 1 cm represents 0.25 km'

71

Do It Yourself

Answers are given in the Answers section at the back of the book.

1 **a** Plot the points given in the following table on a graph:

x	-2	0	2	4	6
y	4	2	0	-2	-4

What is the relationship between x and y?

b (i) Complete the following table for the mapping $x \to x^2$:

x	-2	-1	0	1	2
x^2		1			4

(ii) Draw a graph to illustrate the results from part (i).
(iii) Use your graph from part (ii) to find:
    (a) 1.4^2 (b) $\sqrt{3}$

2 **a** John stood at a radio tower and measured the bearing of the nearby station. He found the answer was 080°. He then walked to the station and measured the bearing of the radio tower from the station. What would he have found as the answer? Explain carefully how you calculated your result.

b A sailing course starts at buoy P. You sail 8 km on a bearing of 010° to buoy Q. You then change course and sail on a bearing of 100°, a distance of 6 km to reach buoy R. Finally, you change course and sail back to P. Draw an accurate diagram of the course, and measure the bearing steered on the final part of the course.

'The scale used depends on the size of the area you want to draw and the size of the page you want to draw it on'

3 A map is drawn to a scale of 1 : 40 000. Find:
(i) the real distance between two stations which are 11 cm apart on the map;
(ii) the distance on the map between two towns that are 20 km apart.

Quadrilaterals, Polygons, Area and Volume

Things You Need to Know

1 The names and properties of different types of quadrilateral (a shape with four sides):

a Rectangle
   (i) All angles are 90°.
   (ii) Opposite sides are equal in length.

b Square
   (i) All angles are 90°.
   (ii) All sides are equal in length.

c Parallelogram
   (i) Opposite sides are equal in length.
   (ii) Opposite sides are parallel.
   (iii) Opposite angles are equal.

d Rhombus
   (i) Opposite angles are equal.
   (ii) All sides are the same length.
   (iii) Opposite sides are parallel.

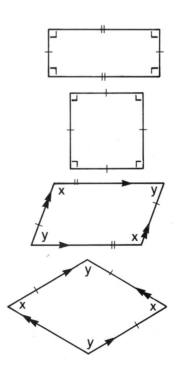

e Trapezium
Two sides are parallel.

f Kite
Two pairs of adjacent sides are equal.

Learn: **The angles of a quadrilateral add up to 360°.**

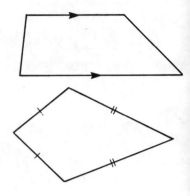

> *All shapes with straight sides, including triangles and quadrilaterals, are known as polygons*

2 Polygons and circles:
A **polygon** has many straight sides.
A **regular polygon** has all sides equal, and all angles equal. Examples include:
a **pentagon**, which has 5 sides
a **hexagon**, which has 6 sides
a **heptagon**, which has 7 sides
an **octagon**, which has 8 sides
A **circle** can be regarded as a regular polygon with an infinite number of sides.
The perimeter, or circumference, of a circle is given by

Perimeter $= \pi d$
(d is the diameter) or $2\pi r$

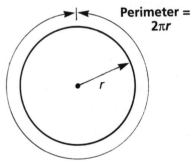

Perimeter = $2\pi r$

3 If shapes can be fitted together so that there are no gaps in between, then they are said to **tesselate**.

4 Area of shapes as follows:
(i) rectangle: area $= ab$

(ii) triangle: area $= \frac{1}{2}bh$

(iii) circle: area $= \pi r^2$

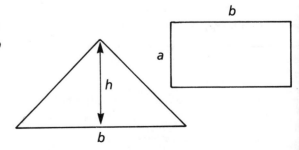

5 Solids

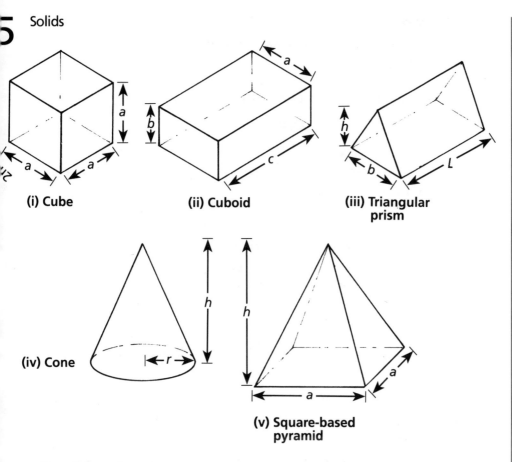

(i) Cube   (ii) Cuboid   (iii) Triangular prism

(iv) Cone   (v) Square-based pyramid

(i) volume of a cube $= a^3$;
(ii) volume of a cuboid $= abc$;
(iii) volume of a triangular prism = area of the triangle × length $= \frac{1}{2}bhL$
(iv) volume of a cone $= \frac{1}{3}$ area of base × height $= \frac{1}{3}\pi r^2 h$
(v) volume of a square-based pyramid $= \frac{1}{3}$ area of base × height $= \frac{1}{3}a^2 h$

How to Do It

1 Referring to the quadrilateral ABCD drawn alongside:
(i) Find x and y.
(ii) What do the answers tell you about the quadrilateral?

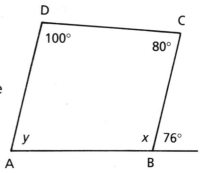

Solution

(i) $x + 76° = 180°$ (angles on a straight line). So

$$x = 104°$$

'*Notice an equation is needed*'

$x + y + 80° + 100° = 360°$ (angles of a quadrilateral). So

$$104° + y + 80° + 100° = 360°$$
$$y = 76°$$

(ii) Since $x + y = 180°$ and also $100° + 80° = 180°$, these are interior angles with AD parallel to CB. So ABCD is a trapezium.

2 Yasmin is trying to work out how many times the wheel of her bicycle will turn when she cycles 50 m across the playground. The radius of the wheel is 25 cm. What will Yasmin find is the answer?

Solution

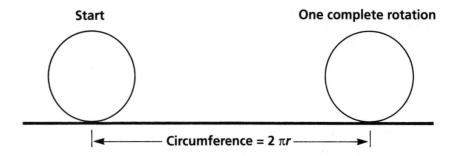

Circumference $= 2\pi r$

The circumference of the circle $(2\pi r)$ is

$$2 \times \pi \times 25\,\text{cm} = 157\,\text{cm}$$

The wheel travels 157 cm $= 1.57$ m while making one complete turn.

'*Always look at one turn for this type of question*'

$$\therefore \quad 50 \div 1.57 = 31.8$$

The wheel makes 31 complete turns.

3 Using only regular octagons of side 2 cm, and squares of side 2 cm, show by using at least four of each type, a pattern that will tesselate.

Solution

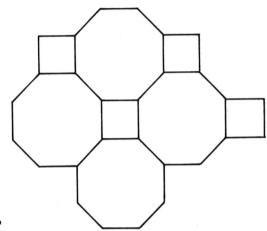

Is this the only one?

4 A piece of string of length 20 cm is first made into the shape of a circle, and then into the shape of a square. Which will have the larger area?

Solution

The sides of the square will be

$$20 \div 4 = 5 \text{ cm}$$

Hence the area is

$$5 \times 5 = 25 \text{ cm}^2$$

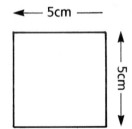

The circumference of a circle is given by
$C = \pi \times$ diameter. Hence the diameter is

$$20 \div \pi = 6.366$$

Which would you expect to contain the larger area?

The radius is $6.366 \div 2 = 3.183$.
   The area of a circle is given by

$$A = \pi \times \text{radius}^2$$
$$= \pi \times 3.183^2$$
$$= 31.8 \text{ cm}^2$$

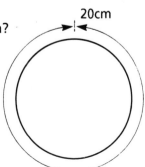

The *circle* has the larger area.

5 The inside tray of a matchbox measures 5 cm by 3 cm by 1.5 cm. A match measures approximately 5 cm by 1.5 mm by 1.5 mm.
- (i) Find the volume of the tray in cubic centimetres (cm³).
- (ii) Find the volume of a match in cubic centimetres.
- (iii) Use your answers to parts (i) and (ii) to find the greatest number of matches that can be fitted into the tray.

Solution

'Make sure you have the units correct'

(i) The volume of the tray is

$$5 \times 3 \times 1.5 = 22.5 \, cm³$$

(ii) The volume of the match is

$$5 \times 0.15 \times 0.15 = 0.1125 \, cm³$$

(iii) The number of matches is

$$22.5 \div 0.1125 = 200$$

Do It Yourself

Answers are given in the Answers section at the back of the book.

1 a Name one fact that makes the following pairs of shapes different:
- (i) a rhombus and a parallelogram;
- (ii) a trapezium and a parallelogram;
- (iii) a square and a kite;
- (iv) a parallelogram and a rectangle.

b Find the value of x in the diagram.

2 a Find the internal angles of:
- (i) a regular pentagon;
- (ii) a regular octagon.

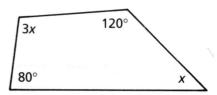

b A coin of radius 1 cm is rolled across the table. If it travels a distance of 80 cm, approximately how many complete turns did the coin roll?

3a Show how to make a tesselation from regular hexagons.

b Show, by using at least five identical kites, how you can tesselate with any kite shape.

4a The diagram on the right shows the outline frame for a door. The top part is semicircular. Find the area taken up by the complete frame.

b Find the area of the bathroom floor shown in the diagram below.

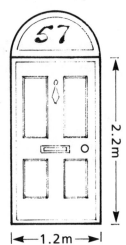

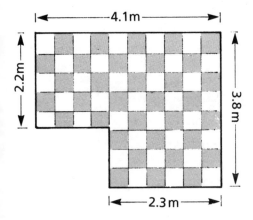

5a The inside of a box measures 1.3 m by 65 cm by 80 cm. Find the volume of the box:

    (i) in cubic metres (m³);

    (ii) in cubic centimetres (cm³);

    (iii) how do you convert cubic metres into cubic centimetres?

b The diagram on the right shows a sketch of a tent. What is the volume of the tent?

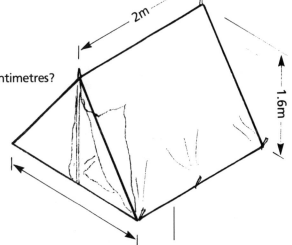

12 | Symmetry, Congruence and Enlargements

Things You Need to Know

1 Shapes can possess different types of symmetry:

(i) **line symmetry** where a shape is exactly the same either side of a **mirror line**, also known as a **line** or **axis of symmetry**

(ii) **rotational symmetry** where a shape is exactly the same after rotation about a point, called the **centre of rotation**. The number of positions that the shape repeats itself is called the **order of rotation**. In the diagram, the shape has an order of rotational symmetry of 3;

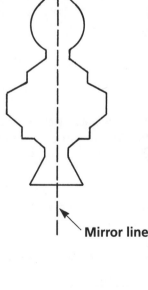

Mirror line

Centre of rotation

(iii) **translational symmetry** where a pattern can be moved in a straight line to repeat itself:

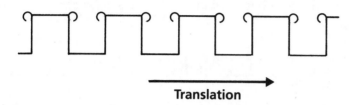

Translation

A shape or pattern may possess more than one type of symmetry.
Two shapes that are exactly the same are said to be **congruent**.

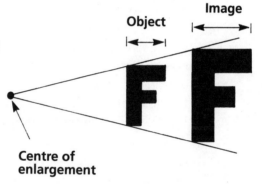

Image

Object

A shape can be magnified by using a mathematical **enlargement**. You need a centre of enlargement, which can be anywhere, and a **scale factor**. This is calculated from the formula

Centre of enlargement

$$\text{Scale factor} = \frac{\text{image length}}{\text{object length}}$$

If the scale factor is a fraction, the shape gets smaller.
If the scale factor is negative, the centre of enlargement is *between* the object and the image, and the image also becomes inverted.

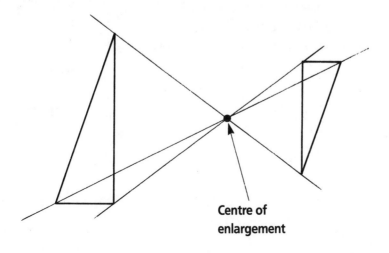

Centre of enlargement

How to Do It

1 **a** For the letters in the following list, state the type of symmetry they possess.

$$A \; B \; C \; H \; L \; M \; O \; Z$$

Where relevant, state the number of axes of symmetry and/or the order of rotational symmetry.

Solution

A   one line of symmetry
B   one line of symmetry
C   one line of symmetry
H   two lines of symmetry and rotational symmetry of order 2
L   no symmetry
M   one line of symmetry
O   two lines of symmetry and rotational symmetry of order 2
Z   rotational symmetry of order 2

'Check with tracing paper if you are unsure'

b This question refers to the 'L' shape drawn on an 8 × 8 grid.
  (i) Rotate the 'L' shape by 180° about the point (5, 4); label the image A.
  (ii) Reflect the 'L' shape in the mirror line; label the image B.
  (iii) Describe how to move position A to position B.

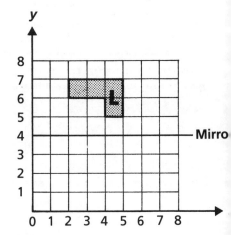

Solution

(i) Check with a piece of tracing paper that L does rotate to position A about the point (5, 4).

(ii) The position of B is shown on the diagram.

(iii) Reflection in the mirror line $x = 5$.

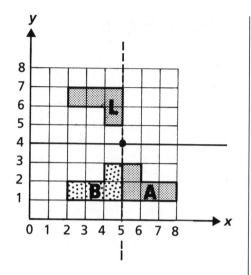

'$x = 5$ is the equation of the mirror line'

2 Draw a grid with x between 0 and 8, and y between -9 and 5. Plot the points A(1, 3), B(4, 3), C(4, 1) and D(1, 1). Enlarge rectangle ABCD by a scale factor 2, with centre of enlargement at (0, 1). Label this A'B'C'D'. Also enlarge ABCD by a scale factor of -2, with centre of enlargement at (3, -1). Label this A''B''C''D''. A'B'C'D' can be transformed to A''B''C''D'' by a rotation. What are the coordinates of the centre of rotation, and what is the angle of rotation?

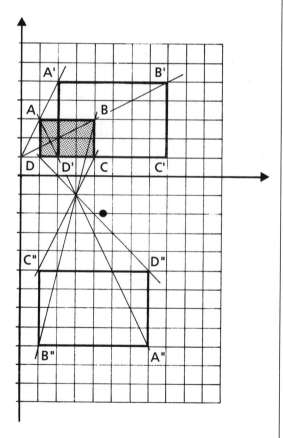

Solution

Follow through the instructions, and check the diagram. To locate the centre of rotation, you may need tracing paper. It is at the point $(4\frac{1}{2}, -2)$. The angle of rotation is 180°.

'D'D'' and C'C'' also meet at the centre of rotation'

Do It Yourself

Answers are given in the Answers section at the back of the book.

1 a For the following shapes, say whether they have line or rotational symmetry. In each case, draw a suitable diagram, showing the lines of symmetry and/or the centre of rotational symmetry. For those shapes with rotational symmetry, state the order of rotational symmetry.

(i) square	(ii) parallelogram	(iii) rhombus
(iv) rectangle	(v) kite	(vi) regular hexagon

b Draw a grid with x between -4 and 4 and y between -4 and 4. Plot the triangle given by A(1, 1), B(1, 2) and C(4, 1) and label it 1. Triangle 1 is rotated by 90° clockwise about the point $(\frac{1}{2}, 1\frac{1}{2})$ to position 2. Draw and label triangle 2. Triangle 2 is then rotated by 90° clockwise about $(0, -2)$ to position 3. Draw and label triangle 3. Describe the single transformation that will take triangle 1 to triangle 3.

'Use tracing paper for this question'

2 a By accurate drawing and measurement, find the centre of enlargement and the scale factor of the enlargement in the following diagram:

'Use a sharp pencil'

b In the space provided, draw an enlargement of the given shape with a scale factor of $-\frac{1}{2}$.

Centre of
enlargement

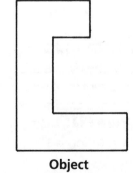

Object

Algebra

13

Things You Need to Know

1 Algebra is a mathematical shorthand. It enables you to work out problems much more quickly. Remember that a letter really stands for a number.

Rules to remember:

(i) $a \times b$ is written ab

(ii) $2 \times a \times b$ is written $2ab$

(iii) $x \times x$ is written x^2

(iv) $a \times x \times x$ is written ax^2

(v) $a \times 0 = 0$

(vi) $6x + 5x = 11x$

(vii) $6x^2 + 5x^2 = 11x^2$

(viii) $4t - 4t = 0$

(ix) $d \div d = 1$

(x) $x \div 4$ can be written $\dfrac{x}{4}$ or $\dfrac{1}{4}x$

(xi) $a \div b$ can be written $\dfrac{a}{b}$

(xii) When different letters are grouped together, this is called **collecting like terms**.

So $4x + 8y + 3x + 2y$ can be grouped to:

$(4x + 3x) + (8y + 2y) = 7x + 10y$

'The key is to get the letter isolated on one side of the equals sign'

2 Equations arise when trying to solve problems. You have to be able to rearrange them to find the value of a given letter.

Some types of rearrangement are:

(i) $$x+5 = 12$$
Subtract 5 from each side $$x = 12 - 5 = 7$$

(ii) $$3x = 12$$
Divide each side by 3 $$x = 12 \div 3 = 4$$

(iii) $$2x - 5 = 6$$
Add 5 to each side $$2x = 6 + 5 = 11$$
Divide each side by 2 $$x = 11 \div 2 = 5.5$$

You should be able to write down an expression using algebra to fit a statement. From these statements you may be asked to produce an equation.

How To Do It

1 a An electrical formula is $P = I^2 R$. Find P if:
(i) $I = 5$ and $R = 20$;
(ii) $I = 0.64$ and $R = 6.38$.

Solution
The formula means $P = I \times I \times R$

(i) $P = 5 \times 5 \times 20 = 500$
(ii) Clearly a calculator is needed here. If you have a square button $\boxed{x^2}$, it can be used as follows:

Display

A sensible answer would be $P = 2.61$.

'Don't leave too many decimal places in the answer'

b A rectangle has sides of length $4x$ and $2y$. Find in its simplest form the perimeter of the rectangle.

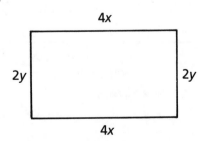

Solution

$$\text{Perimeter} = 4x+4x+2y+2y$$
$$= 8x+4y$$

'Collect like terms'

2a Solve the following equations:

(i) $3x+1 = 10$ (ii) $5(x+2) = 4x+13$ (iii) $8-x = 6-2x$

Solution

(i)
$$3x+1 = 10$$

Subtract 1 from each side $3x+1-1 = 10-1$

So $3x = 9$

Divide each side by 3 $x = 3$

'a(b+c) = ab+ac'

(ii)
$$5(x+2) = 4x+13$$

Remove the bracket $5x+10 = 4x+13$

Subtract $4x$ from each side $5x-4x+10 = 4x-4x+13$

$x+10 = 13$

Subtract 10 from each side $x+10-10 = 13-10$

$x = 3$

(iii)
$$8-x = 6-2x$$

Add $2x$ to each side $8-x+2x = 6-2x+2x$

$8+x = 6$

Subtract 8 from each side $8-8+x = 6-8$

$x = -2$

'You can miss out some steps if you feel confident'

b Jody thinks of a number, doubles it and adds 5. He then trebles the first number and subtracts 7. To his surprise, in each case he arrives at the same number. What number did he think of?

Solution

Try to recognise this type of question as involving an equation.

Let N be the number Jody thought of. To double N and add 5 means $2N+5$, and to treble N and subtract 7 means $3N-7$. If he came to the same answer, then

$$2N+5 = 3N-7$$
$$\text{So} \quad 2N+12 = 3N$$
$$12 = 3N-2N = N$$

The number he thought of was 12.

'You can use any letter unless told otherwise'

c Write an expression for the following:
  (i) the cost of B bolts priced at q pence for 10;
  (ii) the area of a sheet of paper $2x$ cm by $3x$ cm.

Solution

'You are finding the unit cost'

(i) 1 bolt costs $q \div 10$ or $\dfrac{q}{10}$

  B bolts cost $B \times \dfrac{q}{10}$ or $\dfrac{Bq}{10}$

(ii) The area $= 2x \times 3x = 6x^2$.

d Tariq and Samira both attempted to solve the equation $44 - 5x = 3x + 4$. Tariq's answer was $x = 6$ and Samira's answer was $x = 5$. Without actually solving the equation yourself, which answer is correct?

Solution

'Always check your answer'

If you substitute the two answers into the equation, you are not actually solving it; you are checking if either solution is correct. A solution to an equation is a value that makes both sides of the equation equal.

$x = 6$ $44 - 5 \times 6 = 14$ Tariq is wrong
            $3 \times 6 + 4 = 22$

$x = 5$ $44 - 5 \times 5 = 19$ Samira is correct
            $3 \times 5 + 14 = 19$

Do It Yourself

Answers are given in the Answers section at the back of the book.

1 a Simplify as much as possible:

    (i) $4 \times x \times t$ (ii) $6x \div 3$

    (iii) $2x+3x+4x$ (iv) $a \div 2b$

    (v) x^2+x^2 (vi) $4x^2-3x^2$

    (vii) $8t+6q+5t$ (viii) $3q-t+2t$

    (ix) $3x+4y+5x+7y$ (x) $(3x)^2$

b If $a=4$, $b=2$, $t=\frac{1}{2}$, evaluate the following expressions:

    (i) ab (ii) $4at$ (iii) $2a^2$

    (iv) ab^2 (v) a^2+b^2 (vi) $at+bt$

    (vii) $2a^2-b^2$ (viii) abt

    (ix) $a^2b^2t^2$ (x) $2abt^2$

2 a Solve the following equations:

    (i) $x+8=17$ (ii) $2x-5=12$

    (iii) $3x+4=x+11$ (iv) $8-x=15-3x$

b Write down a mathematical expression for the following statements:

    (i) Double the number N and subtract 7.

    (ii) The total cost of x pencils at c pence each.

    (iii) The cost of q oranges which are priced at t pence for a pack of 3.

    (iv) The distance travelled by a car travelling at S miles an hour for t hours.

    (v) The result of adding 8 to N and doubling the answer.

c David is A years old. In 10 years' time, he will be 3 times as old as he was 2 years ago. Write down an equation in A and use it to find out how old David is now.

14 | Decision Trees, Networks and Two-way Tables

Things You Need to Know

1 A **decision tree**, or **flow chart**, is a useful way of writing a set of instructions which occur in succession. It is very often used in designing a computer program.

Instruction box

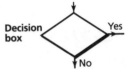
Decision box   Yes   No

2 A **network** is composed of a number of points (called nodes) joined with arcs. They are usually not drawn to scale but just show the relative position of places – for example, the London Underground map.

3 Data can be stored in a table of rows and columns. This is called a **matrix** or **two-way table**. In such an array, the information becomes easier to read.

How to Do It

1 This decision tree is used to generate a number sequence. Work through the tree and write down the number sequence.

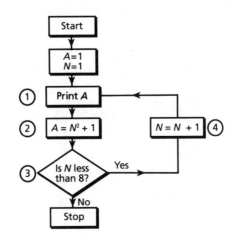

Solution
At ① first time PRINT $A = 1$
At ② first time $A = 1^2 + 1 = 2$
At the decision box ③ $N = 1$ is less than 8 so follow 'yes' route.
At ④ $N = 1 + 1 = 2$
At ① second time PRINT $A = 2$
At ② second time $A = 2^2 + 1 = 5$, etc.
This process continues to give the sequence

$$A = 1, 2, 5, 10, 17, 26, 37, 50$$

'Do not take any short cuts. You must carry out each instruction'

2 The network alongside shows the distances (in miles) between five villages A, B, C, D and E. Joe sets out to cycle from A to D, passing through the other villages only once. How many different routes are there? What is the shortest distance he could cycle?

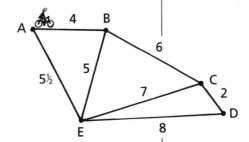

Solution
The possible routes and distances are as follows:

$A \rightarrow B \rightarrow C \rightarrow E \rightarrow D$ $4 + 6 + 7 + 8 = 25$ miles
$A \rightarrow E \rightarrow B \rightarrow C \rightarrow D$ $5\frac{1}{2} + 5 + 6 + 2 = 18\frac{1}{2}$ miles
$A \rightarrow B \rightarrow E \rightarrow C \rightarrow D$ $4 + 5 + 7 + 2 = 18$ miles

Hence there are three routes, and the shortest distance is 18 miles.

'The shape of the network is not important'

3 A bag contains four discs coloured red, green, blue and yellow. Zara removes two discs from the bag one after the other. Use a table to find out in how many ways this can be done.

Solution
The table has the rows and columns labelled by the colours R, G, B, Y.

The number 5 in the table, for example, means that Zara drew green, then blue. It can be seen that there are 12 different ways.

	R	G	B	Y
R		1	2	3
G	4		5	6
B	7	8		9
Y	10	11	12	

Do It Yourself

Answers are given in the Answers section at the back of the book.

1 The numbers {2, 3, 7, 12, 20, 40, 55, 60} are fed into the decision tree shown below.

(i) Enter the answers in the correct box A, B or C.

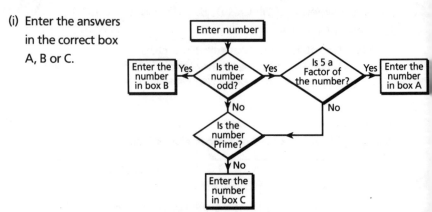

(ii) Give an example of a number that the decision tree cannot sort.

2 Ernie the milkman leaves the head depot (H) and delivers milk at five other depots as shown in the network diagram. (All distances are in miles.) What is the shortest distance he can travel before he has delivered all the milk? If the average speed of his milk float is 6 miles/hour, and the milk must be delivered by 8.00 a.m., what is the latest time he can leave H?

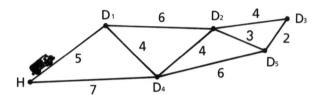

3 The table below shows the milk deliveries to four houses on Friday.

House	Pints	Gold top	Silver top	Red top	Cream
A		2	0	1	1
B		3	1	2	1
C		1	1	0	0
D		2	2	0	0

(i) How many pints of gold top were delivered to the four houses in total?

(ii) How many pints of milk were delivered altogether?

(iii) If the milk costs 34p a pint, and the cream costs 90p a pint, work out how much each house spends on Friday.

Answers

Section 1

1 (i) 1488 (ii) 3010 (iii) 18 647

2 (i) 36 (ii) 112 (iii) 1213

3 **a** (i) 1500 (ii) 430 (iii) 350
(iv) 0.14 (v) 0.7 (vi) 7200

b 192 seeds; £(12 × 320 × 20) ÷ 100 = £768

4 **a** (i) 1, 2, 4, 5, 10, 20
(ii) 1, 2, 4, 5, 8, 10, 20, 40
(iii) 1, 2, 3, 4, 5, 6, 10, 12, 15, 20, 30, 60
(iv) 1, 2, 4, 5, 8, 10, 16, 20, 40, 80

b 8 and 10 go into 40, 40 and 12 go into 120. The smallest number is 120.

5 **a** (i) -6 (ii) $8 \times -2 + 3 - -5 = -8$

b (i) 2 (ii) 36 (iii) -1
(iv) $3\frac{3}{4}$ (v) -7.5 (vi) 12

c $\boxed{5} - \boxed{1} = \boxed{4}$; $\boxed{-3} - \boxed{-4} = \boxed{1}$

6 **a** (i) 300 (ii) 3 (iii) $\frac{3}{100}$
(iv) $\frac{3}{10}$

b (i) $\frac{7}{20}$ (ii) $\frac{27}{40}$ (iii) $\frac{21}{25}$

7 **a** (i) 16 (ii) 243 (iii) 1
(iv) 10 000 (v) 9 (vi) 9

b (i) 4096 (ii) 512 (iii) 2187
(iv) 1.728 (v) 0.000 191 (6 d.p.)
(vi) 35.330 586 (6 d.p.)

c £2 × 12¹² = £2¹³ = £8192

Section 2

1 (i) 16, 22   (ii) $3+5 = 8, 5+8 = 13$
   (iii) 26, 37   (iv) $\frac{31}{32}, \frac{63}{64}$

2 (i) 16, 19   (ii) 61
   (iii) No, because you need 3 links each time to make an extra square; $61 - 3 = 58$, so you would have 2 links left over.

3 **a** (i) 2, 5, 10, 17, 26, 37, 50, 65
    (ii) 2, 7, 12, 17, 22, 27

   b Leaving out the line numbers and the last two lines, you need statements similar to the following:
    (i) FOR N = 1 TO 7 ; PRINT 2*N + 1
    (ii) FOR N = 1 TO 6 ; PRINT N*N
    (iii) FOR N = 1 TO 6 ; PRINT 4*N − 3
    (iv) FOR N = 1 TO 5 ; PRINT N*N*N

Section 3

1 **a**

 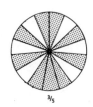 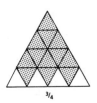

(Different segments can be shaded, so long as the numbers are the same as shown.)

   b 24 cm

   c (i) $1\frac{3}{4}$ (ii) $7\frac{1}{2}$ (iii) $3\frac{4}{5}$

   d (i) $\frac{8}{3}$ (ii) $\frac{13}{4}$ (iii) $\frac{21}{4}$

2 **a** (i) 0.8   (ii) 0.15   (iii) 0.375

   b (i) $\frac{2}{3}$ (ii) $\frac{4}{7}$ (iii) $\frac{3}{4}$

   c (i) $\frac{3}{5}$ (ii) $\frac{6}{25}$ (iii) $\frac{13}{20}$

   d (i) 80%   (ii) 15%   (iii) 24%

3 **a** 28

   b (i) 20%   (ii) 18.75%   (iii) 4%

   c Matthew with 20%

4 **a** $(84 \div 7) \times 2 = 24\,\text{kg}$

   b $(4\frac{1}{2} \div 6) \times 15 = 11.25\,\text{m}$
    Cost $= 80 \times 11.25 = 900\text{p} = £9$

Section 4

1 a (i) 9 (ii) 28 (iii) 0

b (i) 900 (ii) 1000 (iii) 0

c (i) 7 cm (ii) 3 cm (iii) 3 cm

d (i) 6.4 (ii) 8.1
(iii) 1.0 (must have the 0)

e (i) 860 (ii) 6.1 (iii) 39 000

f $1500 \div 65 = 23.07$
Hence 23 glasses can be filled.

g (i) 3 cm (ii) 3 cm (iii) 28 mm

h (i) 38 500 (ii) 38 000 (iii) 40 000

2 a (i) $9 \div 3 = 3$ (ii) $900 \div 300 = 3$
(iii) $100 - 40 = 60$ (iv) $0.1 \times 200 = 20$

b (ii) $7 \times 7 = 49$ (ends in 9) (v) $3 \times 5 = 15$ (ends in 5)
(iii) is also wrong, but you cannot tell just by looking at
the last digit.

3 a 2.84

b 1.64

Section 5

1 (i) 4.64 (ii) 6.3 (iii) 285
(iv) 600 (v) 6.2 (vi) 1250
(vii) 0.26 (viii) 8900

2 (i) 27 (ii) 13.5 (iii) 28
(iv) 35 (v) 224 (vi) 6.25
(vii) 0.25 (viii) 7 ft 4 in

3 a (i) 3.12 in (ii) 78 in

b (i) 48 km (ii) 10.7 km/litre

4 a (i) 3.9 (ii) 22.7 (iii) 33
(iv) 187

b (i) 18 metres/sec (ii) 63 miles/h
(iii) $2\frac{1}{4}$ hours

5 a $87 \times 1\frac{1}{3} = 116$ km (20 minutes is $\frac{1}{3}$ of an hour).

b Time taken $= 2\frac{1}{2}$ hours
Average speed $= 180 \div 2\frac{1}{2} = 72$ km/h

Section 6

a (i) litres   (ii) millimetres   (iii) tonnes
    (iv) metres   (v) kilometres   (vi) minutes

b (i) 40 cm   (ii) 89.2 g   (iii) 1 mile 240 yds
    (iv) 1.6 mm   (v) 100 l   (vi) 2 tonnes
    (vii) 48 minutes

a Radius of the circle = 1.8 ÷ 2 = 0.9 m = 90 cm
    Radius of the bull = 90 − 30 − 40 = 20 cm

b The measurements of the box in centimetres are
    120 × 70 × 60.

$$70 \div 14 = 5$$
$$120 \div 12 = 10$$
$$60 \div 5 = 12$$

    The number of blocks = 5 × 10 × 12 = 600.

Section 7

1 (i) poor: biased   (ii) poor: no time limit given
   (iii) good   (iv) poor: average needed
   (v) probably OK   (vi) good
   (vii) poor: too open-ended   (viii) good

2 Frequencies are 1, 3, 7, 6, 3.

3 (i) Angles for pie chart are: 18°, 54°, 126°, 108°, 54°
   (ii) Bar chart is easier to read accurately.

4 (i)

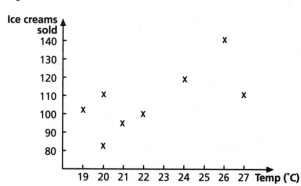

   (ii) No strong correlation, slightly positive.
   (iii) You might expect better correlation.
   (iv) The graph also depends on the day of the week.

Section 8

1 a (i) $\frac{1}{13}$ (ii) $\frac{1}{4}$
  (iii) $\frac{1}{365}$ (ignoring leap years)
  (iv) $\frac{1}{6}$ (v) $\frac{1}{10}$ (vi) $\frac{1}{2}$

b $\frac{7}{25}$

2 a $200 \times \frac{8}{25} = 64$

  b (i) $\frac{2}{5}$ (ii) 2 (on average)

3 a $1 - \frac{5}{9} = \frac{4}{9}$

  b $1 - \frac{1}{5} - \frac{1}{4} = \frac{11}{20}$

4 There are 36 possible outcomes.
  (i) Outcomes for 6 are
    $1+5, 5+1, 2+4, 4+2, 3+3$
    The probability $= \frac{5}{36}$.
  (ii) Scores of 3, 6, 9 and 12 can be obtained as follows:

    $1+2, 2+1, 1+5, 5+1, 2+4, 4+2,$
    $3+3, 3+6, 6+3, 4+5, 5+4, 6+6$

    The probability $= \frac{12}{36} = \frac{1}{3}$.

Section 9

1 (i) 75°   (ii) 60°   (iii) 102°
  (iv) 135°   (v) 30°

2 (i) $a = 70°, b = 70°, c = 70°$
  (ii) $d = 70°, e = 50°, f = 60°$
  (iii) $g = 40° + 60° = 100°$
  (iv) $h = (180° - 120°) + (180° - 110°) = 130°$
      $i = (180° - 100°) + (180° - 130°)$
        $= 80° + 50° = 130°$

3 a (i) 4 cm   (ii) 30°   (iii) 45°

  b (i) Isosceles; angle R $= 72°$, angle Q $= 36°$
    (ii) Equilateral: all sides equal
    (iii) angle R $= 81°$, so the triangle is Scalene.

4

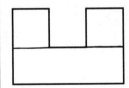 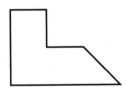

Section 10

a $x + y = 2$

b (i) 4, 1, 0, 1, 4
  (iii) (a) 2 approximately   (b) 1.7 approximately

a $180° + 80° = 260°$

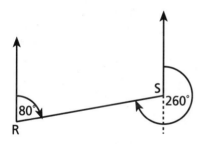

b Final bearing $= 227°$

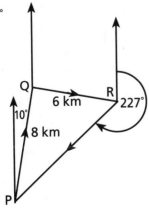

(i) 4.4 km   (ii) 50 cm

Section 11

1 a (i) sides not all equal
  (ii) one pair of opposite sides not parallel
  (iii) one angle not 90°
  (iv) one angle not 90°

b $(360 - 120 - 180) \div 4 = 40°$

2 a (i) $180 - 72 = 108°$
  (ii) $180 - 45 = 135°$

b $80 \div 2\pi = 12$ approximately

3 a

b

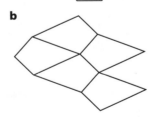

4 a $1.2 \times 2.2 + \frac{1}{2} \times \pi \times 0.6^2 = 3.2 \, \text{m}^2$

b $2.2 \times 4.1 + 2.3 \times (3.8 - 2.2) = 12.7 \, \text{m}^2$

5 a (i) $1.3 \times 0.65 \times 0.8 = 0.676 \, \text{m}^3$
  (ii) $130 \times 65 \times 80 = 676\,000 \, \text{cm}^3$
  (iii) multiply by 1 000 000

b $\frac{1}{2} \times 2 \times 1.6 \times 2 = 3.2 \, \text{m}^3$

Section 12

1 a (i) line and rotational of order 4
   (ii) rotational order 2
   (iii) line and rotational order 2
   (iv) line and rotational order 2
   (v) line
   (vi) line and rotational order 6

b 180° rotation about $(2, -\frac{1}{2})$

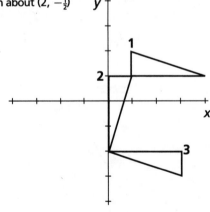

2 a Approximately 2

b

Section 13

1 a (i) $4xt$ (ii) $2x$ (iii) $9x$
   (iv) $\dfrac{a}{2b}$ (v) $2x^2$ (vi) x^2
   (vii) $13t+6q$ (viii) $3q+t$
   (ix) $8x+11y$ (x) $9x^2$

b (i) 8   (ii) 8   (iii) 32
   (iv) 16   (v) 20   (vi) 3
   (vii) 28   (viii) 4   (ix) 16
   (x) 4

2 a (i) 9   (ii) $2x = 17, x = 8.5$
   (iii) $3x-x = 11-4, 2x = 7, x = 3.5$
   (iv) $3x-x = 15-8, 2x = 7, x = 3.5$

b (i) $2N-7$ (ii) xc pence
   (iii) $q \times \dfrac{t}{3} = \dfrac{qt}{3}$ (iv) St miles
   (v) $2(N+8)$ or $2N+16$

c $A+10 = 3(A-2), A+10 = 3A-6$
$10+6 = 3A-A, A = 8$
David is now 8 years old.

Section 14

1 (i) | A: 55 | | B: 2, 3, 7 | | C: 12, 20, 40, 60 |

(ii) The tree cannot sort decimals or fractions.

2 Shortest route $H-D_1-D_4-D_2-D_5-D_3$ gives a distance of 18 miles. This takes 3 hours at 6 miles/hour. Hence he leaves by 5.00 a.m.

3 (i) 8   (ii) 15 (don't include the cream)
(iii) A: £1.92, B: £2.94, C: 68p, D: £1.36

Sample Test Paper

1 Tim is measuring the playground at his school. He measured his own pace to be 76 cm.

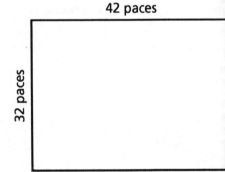

42 paces

32 paces

(i) Without using a calculator and showing full working, find the measurements of the playground:
   (a) in centimetres;
   (b) in metres.

(ii) Estimate the area of the playground in square metres, giving your answer to one significant figure.

(iii) Use a calculator to find the exact value of the area correct to three significant figures.

2

 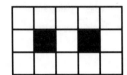 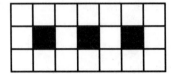

The diagram shows a sequence of patterns made from black and white tiles.

(i) Complete the entries for the next two patterns in the table below:

No. of black tiles	1	2	3		
No. of white tiles	8	13	18		

(ii) Write down in words the rule that connects the number of black tiles (B) to the number of white tiles (W).

(iii) Express this rule in a formula.

(iv) How many black tiles are needed to make a pattern if you have 103 white tiles?

3

Two shops usually sell sweaters at the same price. They both have a sale. Which shop offers the better price? Show all your working.

4 Jane is making teddy bears for the school fête. She found that she could make 6 bears out of a piece of material 2.4 m long.

    (i) How much of the same material would she need to make 9 bears?

    (ii) The material cost £2 per metre. If she can afford to spend £30, how many bears could she make?

5 Katy has been asked to find a solution to the equation $x^3 + x = 40$ by a trial and improvement method.

    1st try: $3 \times 3 \times 3 + 3 = 30$

    2nd try: $4 \times 4 \times 4 + 4 = 68$

    (i) What do these answers tell you about a solution?

    (ii) Continue this method to find the solution correct to one decimal place.

6 A solid has been drawn on isometric paper.

    (i) Draw a plan and elevation of the solid in the direction of the arrow.

    (ii) If the distance between the lines represents 0.5 cm, find the volume of the solid in cubic centimetres.

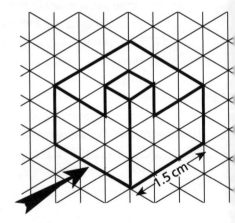

7 What is the probability that a person chosen at random does not like butter?

> **4 out of 7 people prefer butter to margarine**

8 The numbers in ◯ and △ add up to the number in ☐. So

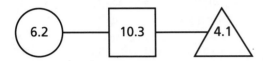

(i) Complete the following:

(a)

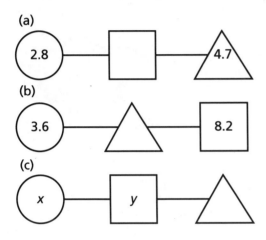

(b)

(c)

(ii) Find x:

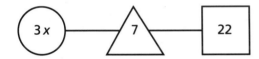

9 Harry, Dave and Pete were working on a building site. Harry worked for 10 hours, Dave for 6 hours and Pete for 9 hours. They earned £375 between them for the job, and agreed to share the money in the ratio of the hours they worked.

    (i) How much did they each earn?

    (ii) Do you think this way of dividing the money is fair? Give a reason for your answer.

10 An aeroplane sets out from an airport (A) and flies on a bearing of 060° for a distance of 80 km to position B. It then changes course and flies on a bearing of 160° for a further 40 km to reach position C. At C it changes course again, and flies in a straight line back to A.

    (i) Draw a scale drawing of the complete journey using a scale of 1 cm to 10 km.

    (ii) What bearing is the plane flying on from C back to A?

11 A cotton reel is 2.2 cm in diameter. A piece of cotton can be wound round the reel 150 times. What is the length (in metres) of the piece of cotton?

12 In the diagram, ABCD is a square. Describe the transformations that will carry out the following moves:

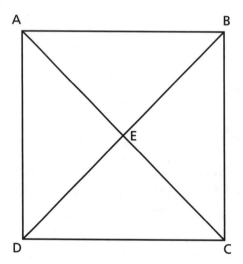

    (i) triangle ABE to triangle ADE;

    (ii) triangle BEC to triangle CED;

    (iii) triangle BEC to triangle DEA.

13 The table below shows the marks scored by seven pupils in music and maths:

	Tina	Albert	Jim	Dan	Bianca	Carol
Maths	40	38	25	18	30	27
Music	26	25	10	15	22	14

(i) On graph paper, plot these data as a scatter graph.

(ii) Comment on the correlation shown.

14 You have been asked to draw up a questionnaire to find out about the types of music that year 9 pupils in your school like. Which of the following questions would be suitable, and which unsuitable? Give a reason for each.

(i) Who is your favourite artist?

(ii) Do you buy CDs or tapes or both?

(iii) How much do you spend each week on music?

(iv) Do you like heavy metal?

(v) You would prefer more pop music on television, wouldn't you?

15 (i) Draw up a table of values for x and y which satisfies the relationship $2x + y = 4$ for y between -6 and 6 and x between -1 and 5. In all cases, x and y must be whole numbers. How many pairs (x, y) can you find?

(ii) Plot the points on a grid. What do you notice?

(iii) Use your graph to find a pair (x, y) which are not whole numbers that satisfy the relationship $2x + y = 4$.

16 The area of a circle is 100 cm². Find the diameter of the circle, giving your answer correct to three significant figures.

Solutions

1 (i) (a)

$$\begin{array}{r} 76 \\ \times 32 \\ \hline 2280 \\ 152 \\ \hline 2432 \end{array}$$

and

$$\begin{array}{r} 76 \\ \times 42 \\ \hline 3040 \\ 152 \\ \hline 3192 \end{array}$$

The measurements are 3192 cm by 2432 cm.

(b) 31.92 m by 24.32 m.

(ii) $30 \times 20 = 600\,\text{m}^2$

(iii) $31.92 \times 24.32 = 776\,\text{m}^2$ (3 s.f.)

2 (i)

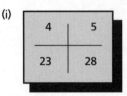

4	5
23	28

(ii) To say that W increases by 5 would not be sufficient. To find the number of white tiles, you multiply the number of black tiles by 5 and add 3.

(iii) $W = 5B + 3$

(iv) $5B + 3 = 103$ gives $B = 20$. Hence 20 black tiles are needed.

3

$$\tfrac{1}{3} = 1 \div 3 = 0.33$$

$$30\% = \frac{30}{100} = 0.3$$

Hence the greater reduction is by Zigzag. This means Zigzag offers the better price.

4 (i) 1 bear needs $2.4 \div 6 = 0.4$ m.
Hence 9 bears need $9 \times 0.4 = 3.6$ m.

(ii) £30 would buy 15 m.

$$15 \div 0.4 = 37.5$$

Jane can make 37 bears (not 38).

5 (i) A solution lies between 3 and 4.

(ii)

3rd try:	$3.5^3 + 3.5 = 46.375$	too high
4th try:	$3.4^3 + 3.4 = 42.704$	too high
5th try:	$3.3^3 + 3.3 = 39.237$	too low

The nearest answer is 3.3.

6 (i)

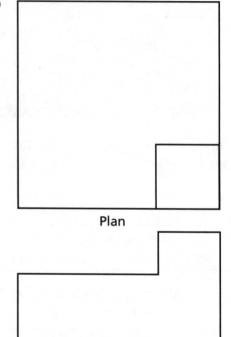

Plan

Elevation

(ii) The shape consists of a cuboid and a cube.

Volume of cuboid $= 1.5 \times 1.5 \times 1 = 2.25\,\text{cm}^2$
Volume of cube $= 0.5 \times 0.5 \times 0.5 = 0.125\,\text{cm}$
Total volume $= 2.375\,\text{cm}^3$

7

$$1 - \frac{4}{7} = \frac{3}{7}$$

8 (i) (a) $2.8 + 4.7 = 7.5$

(b) $8.2 - 3.6 = 4.6$

(c) $y - x$

(ii) $3x + 7 = 22$ so $3x = 15$ and $x = 5$

9 (i)

Total hours $= 10 + 6 + 9 = 25$
£375 $\div$ 25 $=$ £15

So

Harry earns $10 \times £15 = £150$
Dave earns $6 \times £15 = £90$
Pete earns $9 \times £15 = £135$

(ii) The method is fair because it is exactly the same as finding out how much they are being paid per hour.

10

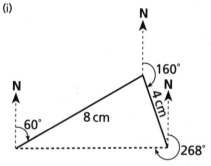

(i)

(ii) Bearing = 268° (±2°)

11 The circumference of the circle = $\pi \times 2.2 = 6.91$ cm
The length of cotton = $150 \times 6.91 = 1036.5$ cm
Divide by 100 to give 10.365 m.

12 (i) reflection in AE;
(ii) rotation of 90° clockwise about E;
(iii) rotation of 180° about E.

13 (i)

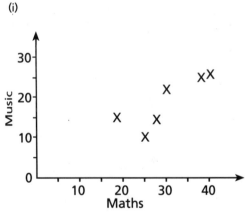

(ii) The graph shows some positive correlation.

14 (i) Not suitable: too many possible answers.
(ii) Suitable: has a definite answer.
(iii) Suitable if a choice given; £0–2, £3–4 for example.
(iv) Suitable: has a definite answer.
(v) Not suitable: it is a leading question.

15 (i)

x	−1	0	1	2	3	4	5
y	6	4	2	0	−2	−4	−6

These are the only pairs. Hence there are six points.

(ii) The grid is shown below. The points all lie on a straight line.

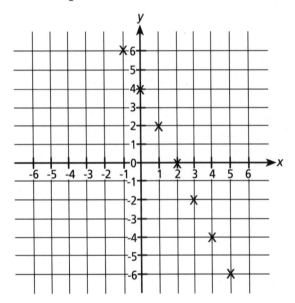

(iii) There are many possibilities: for example, (1.4, 1.2).

16 $$\pi r^2 = 100$$
So $$r^2 = 100 \div \pi = 31.831$$
Hence $$r = \sqrt{31.831} = 5.64$$
Diameter $d = 2r = 11.3$ (3 s.f.)

Index

Titles in *The Way to Pass* series

These books are available at £7.99 each from all good bookshops or directly from Vermilion (post and packing free) using the form below, or on our credit card hotline on **0279 427203**.

ORDER FORM

National Curriculum Maths

				Quantity
Level 4	Key Stage 3	11-14 years	0 09 178116 7	....
Level 5	Key Stage 3	11-14 years	0 09 178118 3	....
Level 6	Key Stage 3	11-14 years	0 09 178125 6	....
GCSE Foundation Level	Key Stage 4	14-16 years	0 09 178123 X	....
GCSE Intermediate Level	Key Stage 4	14-16 years	0 09 178121 3	....
GCSE Higher Level	Key Stage 4	14-16 years	0 09 178127 2	....

National Curriculum English

Level 4	Key Stage 3	11-14 years	0 09 178129 9	....
Level 5	Key Stage 3	11-14 years	0 09 178135 3	....
Level 6	Key Stage 3	11-14 years	0 09 178133 7	....
GCSE	Key Stage 4	14-16 years	0 09 178131 0	....

Mr/Ms/Mrs/Miss..

Address:...

....

....

Postcode:.. Signed:..

HOW TO PAY

I enclose cheque / postal order for £........ : ....made payable to VERMILION
I wish to pay by Access / Visa card (delete where appropriate)

Card No ....Expiry date:...............................

Post order to **Murlyn Services Ltd, PO Box 50, Harlow, Essex CM17 ODZ.**

POSTAGE AND PACKING ARE FREE. Offer open in Great Britain including Northern Ireland. Books should arrive less than 28 days after we receive your order; they are subject to availability at time of ordering. If not entirely satisfied return in the same packaging and condition as received with a covering letter within 7 days. Vermilion books are available from all good booksellers

The Video Class Mathematics and *English* videos which accompany the above titles are available at £12.99 from leading video retailers and bookshops, or on the credit card hotline **0275 857017.**